Microclimate Measurement for Ecologists

Biological Techniques Series

J. E. TREHERNE
Department of Zoology
University of Cambridge
England

P. H. RUBERY
Department of Biochemistry
University of Cambridge
England

---

1. Ion-sensitive Intracellular Microelectrodes, *R. C. Thomas* 1978
2. Time-lapse Cinemicroscopy, *P. N. Riddle* 1979
3. Microclimate Measurement for Ecologists, *D. M. Unwin* 1980.

Microclimate Measurement for Ecologists

D. M. Unwin

Department of Zoology
University of Cambridge
Cambridge CB2 3EJ

1980

ACADEMIC PRESS
London·New York·Toronto·Sydney·San Francisco
A Subsidiary of Harcourt Brace Jovanovich, Publishers

ACADEMIC PRESS INC. (LONDON) LTD
24 – 28 Oval Road,
London NW1

U.S. Edition published by
ACADEMIC PRESS INC.
111 Fifth Avenue
New York, New York 10003

British Library Cataloguing in Publication Data

Unwin, D
Microclimate measurement for ecologists.
(Biological techniques series).
1. Microclimatology—Measurement
I. Title II. Series
551.6'6'024574 QH543 80-41246

ISBN 0-12-709150-5

PRINTED IN GREAT BRITAIN BY
JOHN WRIGHT & SONS LTD, AT THE STONEBRIDGE PRESS, BRISTOL

Foreword

Almost all scientific discovery today depends on equipment and unfortunately this means that in most cases research is limited to those with advanced training and access to very expensive facilities. Happily, however, there is one large area of real importance to man where this need not be true, and where there are thousands of situations which need investigation. This is the understanding of the relationship between living things and their physical environment, whether this be in the natural or in an agricultural situation.

It has been known for a long time that the life of most organisms is controlled by large-scale changes: of the seasons, drought and so on. But after attempting for years without great success to demonstrate the relationship between events, like the pollination of plants or the behaviour of insects, with the "weather" as measured on a large scale by ordinary meteorological instruments, we are beginning to understand that what really matters is the climate within a couple of millimeters of a leaf surface, or actually inside a flower or a small hole in the soil; these climates can be entirely different to the weather grossly measured outside. Sometimes the plant or animal will create a miniature climate for itself. The humidity immediately around a small object will determine whether it will dry up rapidly. Taking up or losing heat is essential to the energy balance of many creatures. The level of light is often the trigger for key pieces of behaviour.

There are many thousands of plants and animals, each with its own special requirements, each one different or they would not be competing and surviving species; while especially in the agricultural context there are many varieties within a single species. A huge amount of good science is there to be achieved by a combination of keen observation and relatively simple instruments which can be taken into the field.

Dennis Unwin has spent many years discussing with biologists their problems about things they want to measure. He has devised and constructed equipment to solve those problems. He knows the difficulties many biologists have in clarifying their understanding of physics, and what it is practical to expect a biologist to be able to construct and to use in the field as well as the laboratory. He happens also to be an able dipterist who has used the machines described in this book in field investigations. Many of them can be built inexpensively in a school or college workshop. Used in conjunction with his lucid explanation of the physical factors of climate and the microenvironment, this should make possible a range of investigation in the field and on expeditions, of immense importance to ecology. I believe this book will be unusually valuable to pure and applied field biologists of all levels.

Sir James Beament Sc.D., F.R.S.,
Drapers Professor of Agriculture in the University of Cambridge;
Chairman of the Natural Environment Research Council.

Preface

This book is intended for biologists who need to measure the very local climate in which plants and animals live and do not wish to become microclimatologists in order to do so. Microclimatology (or micrometeorology) is the study of the climate in the boundary layer of the atmosphere, where factors such as temperature and humidity can change dramatically in a distance of a few centimetres and where plants and animals can modify the climate in which they and other organisms live. Meteorological instruments of the type used in weather stations are generally unsuitable for this scale of measurement but a wide range of specialised microclimatic instruments is available to the ecologist (Monteith, 1972) and it is one function of this book to explain how they work and how they are used. It is also possible to build simple but effective instruments, which do not require a high level of expertise in their construction, and many such designs are included.

The first four chapters deal with the measurement of temperature, radiation, humidity and wind respectively. Each of these chapters includes sections on the principles of measurement, how instruments work, possible sources of error in measurement, detailed descriptions of simple instruments that can be made by the experimenter and procedures for the testing and calibration of instruments. The fifth chapter is concerned with miscellaneous topics including the control of temperature and humidity on a small scale, artificial lighting and how to make a remotely controlled microbalance for investigating changes of weight inside a sealed chamber. The final chapter is concerned with humidity calculations and includes a set of humidity tables. Suggestions on how to build electronic circuits and where to obtain supplies of components and materials are given in the appendix.

It is likely that many ecologists who involve themselves in the measurement of microclimate will wish to learn more of the science of micrometeorology. Some excellent books have been written on the subject: Geiger (1965), Monteith (1973), Rosenberg (1974) and Schwerdtfeger (1967) call for particular mention.

Finally, I would like to mention some friends who have helped me to write this book. Although I had been involved with microclimate instrumentation for a long period, it was Dr Sarah A. Corbet, of the Department of Applied Biology, Cambridge, who involved me in a research project which provided the stimulus to develop some new techniques. Without her influence, this book would never have started. I am also greatly indebted to my colleague, Dr P.G. Willmer, who volunteered to read the manuscript and made many valuable comments and suggestions.

July, 1980 D. M. Unwin

Contents

3
Humidity

4
Wind

5
Miscellaneous Topics

1
Temperature

I. The Measurement of Temperature

The measurement of temperature is important in its own right as well as being used in the determination of other microclimate parameters, and a wide range of techniques has been developed. Since the physical properties of most materials vary with temperature, it is inevitable that many more techniques will be developed in the future. With the exception of radiation thermometers, which have problems of their own, all that a temperature indicator tells you is the temperature of its sensor and it is for the users to decide how this relates to the temperature that they wish to measure. A large metal-sheathed resistance thermometer calibrated to 100th of a degree may give a less accurate estimate of the surface temperature of soil than a small thermocouple that has been calibrated to half a degree. Generally, the smaller the sensor the easier it is to use. Physical effects that are useful in measuring temperature fall into six main groups as follows.

A. *Expansion of Solids, Liquids or Gases*

Thermometers in which liquids like mercury, alcohol or toluene are contained in a glass or metal bulb connected to a capillary that indicates changes in volume are well known. Such a bulb may also be connected to a pressure gauge or recorder and indicate a change in pressure. The change in volume or pressure of gases may also be used for indicating temperature. Thermometers using solid materials normally compare the length of two materials that have different coefficients of linear expansion and use levers and other mechanical devices to magnify the relative changes in length. Sometimes bimetallic strips are used; these are made from two metals with very different coefficients of expansion which are soldered together, and the curvature of the strip is a function of temperature. This is the method used in the well-known clockwork thermohydrograph. Instruments in this group are generally unsuitable for the use of remote indicators. Mercury bulbs may be connected to pressure gauges through narrow-bore steel tubing, but the distances involved can only be a few metres. For remote indication, electrical methods are very much more convenient. The expansion type of thermometer also tends to have a large sensing element with high thermal mass and may be slow in response. It must also be said that there is very little that can go wrong with them and that, if they do fail, they usually do so in an obvious way that is immediately noticed. For this reason, expansion thermometers are often used as standards.

B. *Change of Electrical Resistance*

In true metals, resistance increases with increase in temperature. In the semiconductor germanium and the electrically conducting non-metal carbon, resistance decreases with increase in temperature. These effects can be used for measuring temperature, and many commercial thermometers have been built using this principle. In some older instruments, coils of copper wire are used as the sensor but these seem to have been superseded by the platinum resistance element. In this device, a thin film of platinum is deposited on an insulated substrate and forms a sensor that has a conveniently high resistance and quite small physical size. A Wheatstone bridge can be used to measure the resistance of the element but it is more usual now to pass a constant current through the sensor and measure the voltage. With metal resistance thermometers the relationship between resistance and temperature may be considered to be linear over the temperature range used by ecologists. The main source of error is corrosion of the element and for this reason the platinum element is to be preferred although non-precious metal sensors can be protected. The use of stainless steel is perhaps worth investigating. Of the group of resistance thermometer elements in which the resistance decreases with increase in temperature, the most useful device is the thermistor. Thermistors are usually very much more sensitive to changes in temperature than metal resistance elements, although they are not as stable and the relationship between temperature and resistance is not linear. They can be very small and the non-linearity can be largely compensated for in the circuitry so that, despite their slight lack of stability, they are one of the most commonly used temperature sensing elements. Modern thermistors are much more stable than earlier types and provided large thermal shock is avoided, they will keep their calibration over a long period.

In order to measure the resistance of the sensor some current has to be passed through it and some power must be dissipated in the resistance element. This is particularly important when the temperature of still air is being measured and will result in the sensor indicating a value slightly above air temperature. There is no way of allowing for this increase because its value will depend on the air movement past the sensor, and the best policy is to keep the power dissipated in the sensor to a minimum consistent with accurate measurement. A method of checking for this effect will be described later in this chapter.

C. Change in Electrical Potential

There are two methods of measurement in this group, the silicon diode and the thermocouple. The forward voltage of a silicon diode decreases by about 2 millivolts (mV) per degree Celsius (°C) rise in temperature. (The forward voltage is the voltage developed across the diode when current is passed in a forward direction; its normal value is about 600 mV at room temperature and also depends on the current.) The relationship between voltage and current is stable and linear, and the device costs much less than a thermistor. The smallest silicon diodes are larger than the smallest thermistors and are effectively rather less sensitive, but their good linearity and low cost make them a very useful device.

The thermocouple, which is made from two dissimilar metals, produces a voltage that is proportional to the difference in temperature between the two junctions. The voltage produced is very small (about 38 μV/°C for copper and constantan junctions) but the thermocouple has three very good points in its favour. It can measure differences in temperature, it has very good linearity and it can be made very small, down to a few microns (micrometres μm) if necessary. Unlike other electrical sensors, self-heating is not a problem.

D. Radiation from a Surface

Since the wavelength of maximum emission of any black body radiator is inversely proportional to its absolute temperature (Wien's displacement law), we can use sensors in the infra-red region to measure surface radiation (around 10 μm wavelength) and distinguish it from the sun's radiation reflected from the surface (0.48 μm wavelength). The intensity of radiation received by such a sensor will be proportional to the fourth power of the absolute temperature of the surface. Because it is a fourth power function, the radiation changes considerably over the biological range of temperatures and sensors have been mounted in aeroplanes and satellites that can measure very small changes in the earth surface temperature. The estimation of surface temperature by this method makes assumptions about the emissivity of the surface and results should therefore be interpreted with care, but it seems likely that increasing use will be made of this method on a small scale. Surface temperature radiometers can be obtained which will estimate the average temperature of a surface from a distance, and although they generally have a rather wide angle of acceptance (about 7 degrees), the resolution is expected to be improved in future designs.

E. Change in Resonant Frequency

The frequency of oscillation (and hence the time keeping) of old pendulum clocks was temperature dependent because the pendulum expanded and contracted with changes in temperature. Much the same thing happens to the resonant frequency of a quartz crystal although the temperature coefficient of a crystal that has been specially cut for standard frequency applications will be extremely small. However, crystals can be cut that have much higher temperature coefficients and these can be used for the very precise measurement of temperature.

F. Chemical Temperature Integration

This method of estimating mean temperature uses the fact that the velocity of a chemical reaction varies with temperature, so that if a slow and irreversible chemical reaction is subjected to a fluctuating temperature, the extent to which the reaction has proceeded gives a measure of the mean temperature. The most commonly used reaction is the hydrolysis of sucrose, and an assessment of the technique is given by Lee (1969). The sensing elements are 25 ml ampoules filled with a concentrated sucrose solution, with small amounts of sodium citrate and hydrochloric acid, and the degree of inversion of the sucrose is indicated by the optical rotation, measured in a polarimeter. This technique has the considerable advantage that the sensors are simple and inexpensive. Its main disadvantage is that the response of sucrose hydrolysis to temperature changes is exponential, and if the range of temperatures to which the sensor is exposed is large, this can result in serious errors. In general, this is a useful method for estimating the mean temperature of damp soil or water bodies, where the range of temperature is low, but is generally unsatisfactory for the measurement of mean air temperatures near the ground.

II. Major Sources of Error in Temperature Measurement

The two most important sources of error in temperature that are not just shortcomings of the measurement system are concerned with the coupling of the sensor with its environment. These are the effects of incoming radiation on the temperature of the sensor and the effect of the thermal mass of the sensor. Both effects become more important in air than in water, because of the lower thermal coupling between the sensor and the environmental medium.

Thermal inertia depends on the thermal mass of the sensor and the medium being sensed. A large sensor in air will not respond to short-term variations in air temperature. Whether this matters will depend on the type of measurement being undertaken. In a portable instrument it is inconvenient to have to wait a considerable time for a reading and a response time of two or three seconds is usually about right. Too fast a response makes the thermometer difficult to read and should be avoided unless there is a good reason for requiring an instantaneous reading. If a large number of sensors are connected to a data logging system, the response time of the sensors should be comparable with the sampling interval if this is reasonably rapid; if sampling is infrequent, then the response time should be something like a minute so that short-term variations, which the system cannot record, are excluded.

The effect of radiation on temperature sensors is generally the most serious source of error in temperature measurement. If the sensor of a thermometer is exposed to the sun's radiation, a fraction of that radiation, determined by the short-wave absorptivity of the surface, will be absorbed and the remainder reflected away. Since objects nearby are also reflecting the sun's radiation, the sensor will receive short-wave radiation reflected from them. It will also receive long-wave radiation radiated from nearby objects and it will itself emit long-wave radiation. It is therefore not surprising that the temperature of the sensor will be different from that of the air. If the total incoming radiation exceeds the outgoing radiation (the net radiation is positive) as generally happens during the day, the sensor will be above air temperature. When the total incoming radiation is less than the outgoing radiation (net radiation negative), a situation that usually occurs at night but may occur during the day in open shade (Stoutjesdijk, 1974), then the sensor will be below air temperature. The magnitude of this difference will depend on the surface of the sensor, whether any shielding has been provided, and the amount of air movement. Meteorological instruments are generally placed in white-painted louvred boxes that are intended to shield the instruments from radiation and allow natural ventilation, but under very calm conditions considerable errors can occur. When very accurate measurements are required, ventilated shields are used in which air is drawn through the radiation shield and past the sensor by an electrically driven fan. If a large sensor, such as a mercury-in-glass thermometer, is to be used to measure air temperature, then it is worth taking two simple precautions: the surface absorptivity

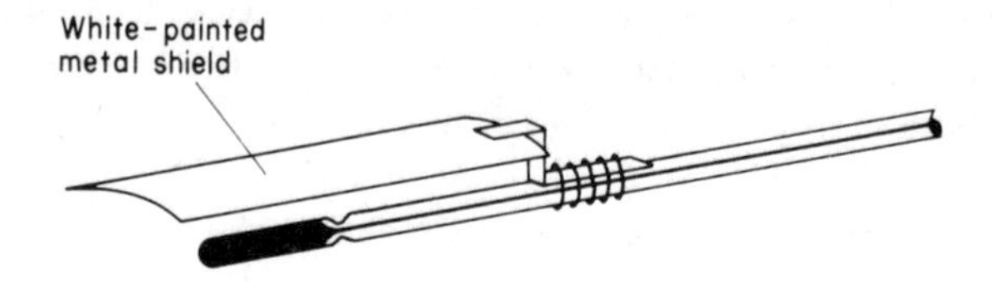

Fig. 1. Radiation shield for a mercury thermometer.

can be reduced by painting the bulb with a good quality white gloss paint, and it can be fitted with a radiation shield (Fig. 1). Provided there is some air movement, it will then indicate a reasonable approximation to air temperature. The size of the sensor also has an effect on the accuracy of measurement when used under conditions where natural ventilation is relied upon. Very small sensors only need a very small volume of air moved past them in order to reduce the temperature difference between the air and the sensor to acceptable limits.

III. Simple Equipment for Temperature Measurement

A. *Thermistor Temperature Meter*

The relationship between temperature and resistance for a thermistor is shown in Fig. 2. The fact that this relationship is a curve rather than a straight

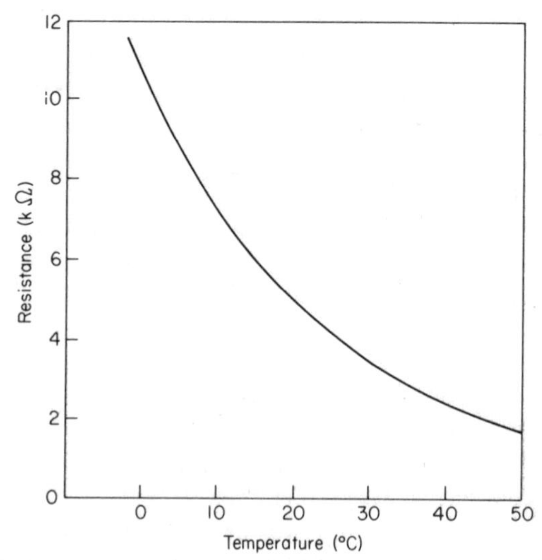

Fig. 2. The relationship between temperature and resistance for the thermistor used in Figs. 3 and 4.

line is the main disadvantage of using thermistors. It is possible simply to calibrate the meter so that this relationship is taken into account but it is generally more convenient to use a linear scale, and this can be achieved by connecting a resistor of appropriate value in parallel with the thermistor. The value of the resistor (R in Figs 3 and 4) will depend on the characteristics of the thermistor and the compensation is not perfect, but with care a thermistor can be made linear to within ± 0.1°C over the range 0 – 30°C and to within ± 0.25°C over the range 0 – 40°C. Outside this range, serious errors occur and if wide range measurements are required, then consideration should be given to other sensors, such as a silicon diode or a metallic resistance thermometer. If a non-linear scale can be tolerated, there is no reason why a thermistor should not be used over a wide temperature range. Although thermistors are very sensitive, amplification is needed in order to drive a meter and this can be provided by a simple integrated-circuit operational amplifier, as shown in Fig. 3. Amplification will not be needed, however, if a sensitive indicator such as a digital panel meter is used, and this is the approach used in the design shown in Fig. 4.

The measurement circuits of the two designs shown in Figs 3 and 4 are identical, with the thermistor (T) and its compensating resistor (R) forming one of the arms of a bridge. The bridge is in balance at 0°C, this setting determined by adjustment of the multi-turn pre-set potentiometer Q. The maximum reading on the meter scale, in the case of an analogue meter, or the high temperature setting point in the digital design, will be determined by the

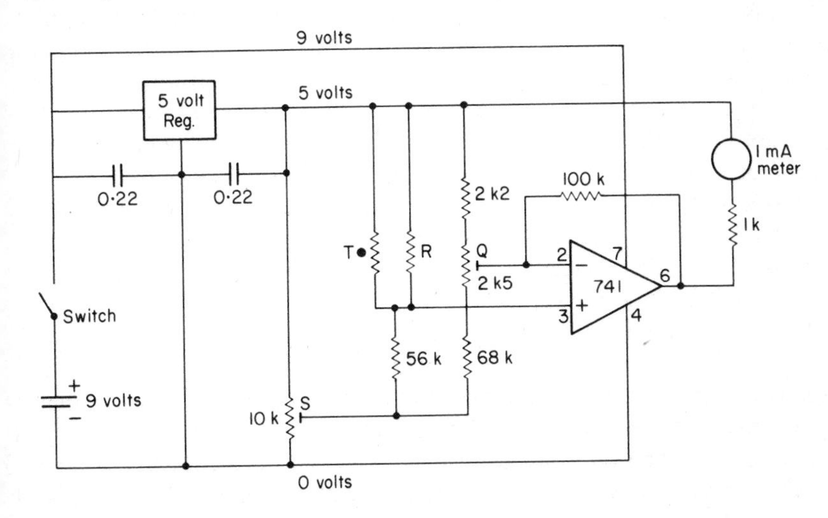

Fig. 3. Circuit diagram of the thermistor meter with analogue output.

T is a 4k7 thermistor type GM472 or VA3404 (R.S.Components Ltd, 151 – 142).

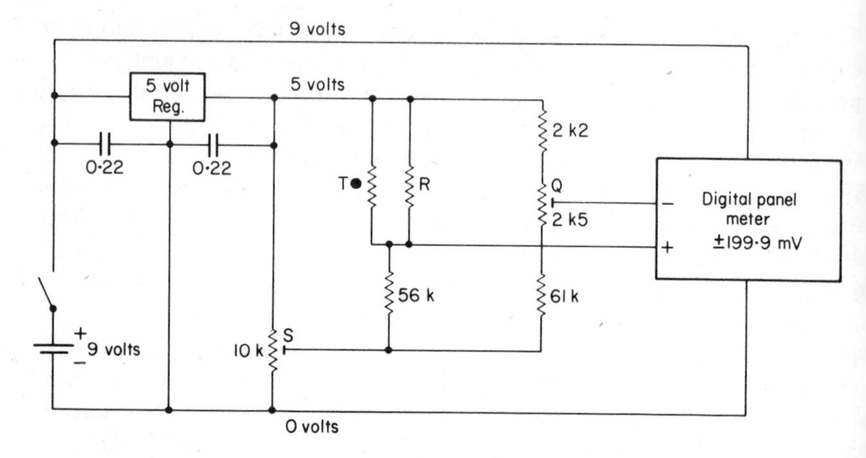

Fig. 4. Circuit diagram of the thermistor meter with digital output.

T is a 4k7 thermistor type GM472 or VA3404 (R.S.Components Ltd, 151 – 142).

setting of the bridge voltage potentiometer (S). It is suggested that either 30 or 40°C should be chosen as the high temperature setting point, depending on the range and accuracy required. The compensating resistor (R) should be selected for accurate reading at the mid-point of the range. For the thermistor used in these designs, the value was found to be about 4000 Ω. If the upper limit and the 0°C points have been set and it is found that the meter reads high at the mid-point of the range, R should be reduced. When the extremes and the mid-point of the range are all accurately set, deviation from linearity will be found to be S-shaped, reading lowest at one-quarter and highest at three-quarters of the range. The exact amount of power dissipated in the thermistor will depend on the characteristics of the individual thermistor and the temperature being measured but is in the region of 1 microwatt (μW). This will result in a very small error due to self-heating of the thermistor: further consideration will be given to this problem in the section on testing and calibration later in this chapter.

It may be thought that the use of a digital panel meter, reading to one-tenth of a degree is an unnecessary luxury in this type of equipment but this is not really the case. The cost of the digital panel meter is about 50% greater than a good analogue meter at present, but this gap is narrowing and relative cost will soon cease to be a consideration. A liquid crystal device uses about 1 milliamp (mA), rather less than an analogue meter and operational amplifier, and is more robust. With the digital meter, the system has a precision of about one-tenth of a degree but an accuracy of perhaps only a quarter of a degree (if a 40 degree range has been chosen). This does not mean that this precision is ot useful since it is often more important to be able to measure the difference etween two similar temperatures than to know their absolute value.

It is possible to obtain precision compensated thermistors, which are ther- ıistors with appropriate compensating resistors built-in. They are an order f magnitude more expensive than simple thermistors, and usually larger, ınce the bulb contains a resistor as well as the thermistor, and whether they re worth using will depend on the application. If they are used in the circuits f Figs 3 and 4, the resistor R should be omitted, and it may be necessary to lter the value of the 2.2 kilohm resistor adjacent to the potentiometer, Q. A ompensated thermistor with a resistance of about 2000 Ω would be expected) perform satisfactorily in the circuits of Figs 3 and 4 without modification.

B. Silicon Diode Temperature Meter

he relationship between temperature and the forward voltage of a silicon iode is almost a straight line, and over a range of 0 – 40°C deviates only bout 0.1°C from linearity. Outside this range, errors gradually increase and meter that is adjusted to read correctly at 0 and 40°C will over-read by bout 1°C at 70°C. However, the excellent performance of the silicon diode t the temperatures normally encountered in microclimate work makes it a ery useful device. Silicon diodes are less sensitive than thermistors and gain, amplification is needed.

The circuit diagram of an analogue silicon diode temperature is shown Fig. 5. A small current is passed through the diode (D) and the voltage

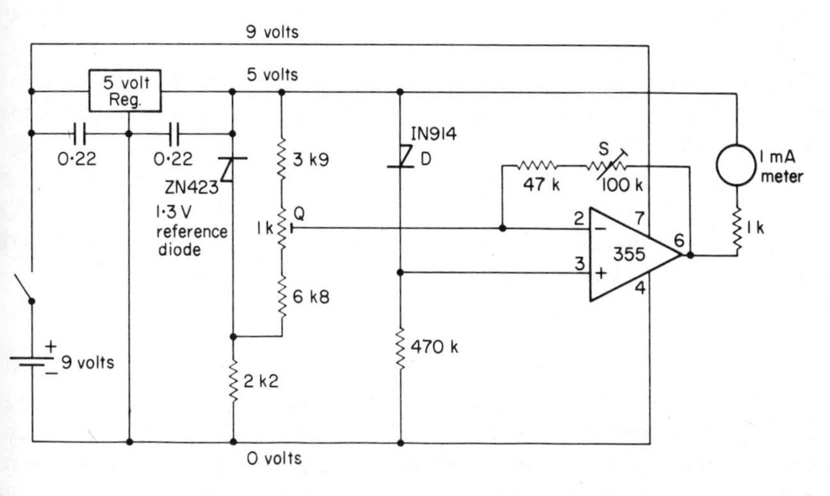

ig. 5. Circuit diagram of the silicon diode temperature meter with analogue output. D is a 1N914 silicon diode used as the sensor (R.S.Components Ltd, 271 – 606).

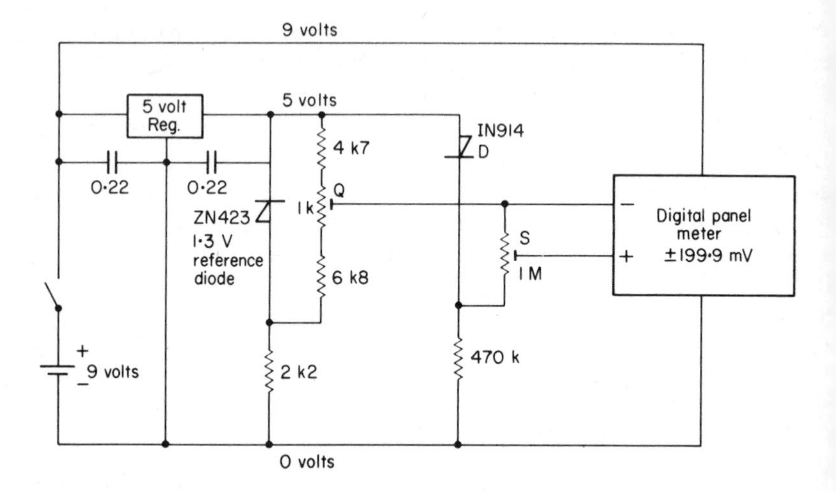

Fig. 6. Circuit diagram of the silicon diode temperature meter with digital output. D is a 1N914 silicon diode used as the sensor (R.S.Components Ltd, 271 – 606).

developed is compared with a stabilised voltage by an operational amplifier In order to keep the power dissipated in the measurement diode (D) to the microwatt level, the resistance feeding the diode must be high and the input impedance of the amplifier has to be even higher, requiring the use of an FET (field-effect transistor) operational amplifier. The multi-turn pre-set potentiometer (Q), which is fed from a 1.3 volt precision reference diode, is adjusted for zero output from the amplifier at 0°C. The feedback resistor (S) is adjusted for correct reading at 30 or 40°C.

Figure 6 shows the alternative version of the circuit with a digital panel meter replacing the analogue meter and amplifier. Once again, the potentiometer (Q) sets the zero point and the gain control (S) sets the 40°C calibration. Some digital panel meters have provision for adjustment of the sensitivity and may have a range large enough to make the potentiometer (S) unnecessary. In this case, the plus input of the meter should be connected directly to the junction between the diode (D) and the 470k resistor.

C. *Thermocouples*

The thermocouple (Fig. 7) is a useful and versatile device for measuring temperature. It consists of a pair of junctions between dissimilar metals, such as copper and constantan alloy, and the electrical output measured between the two copper leads is approximately proportional to the difference in temperature between the two junctions (T1 and T2). The voltage produced by

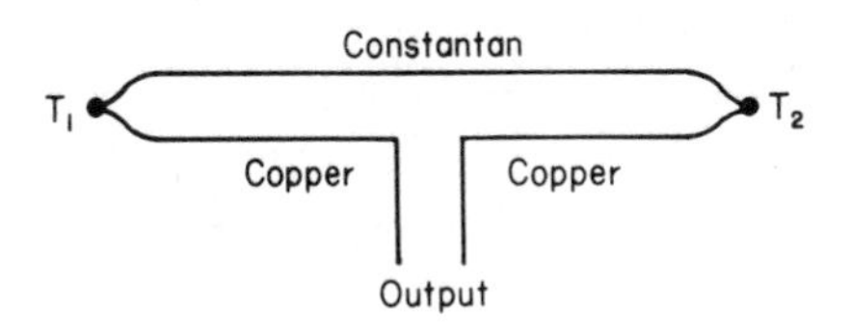

Fig. 7. A pair of thermocouples.

thermocouples is quite small, and slightly dependent on the temperature at which the difference in temperature is measured. A difference of 1°C in temperature between the junctions of a copper – constantan thermocouple will result in an output of 38 μV at 0°C, 42 μV at 40°C and 46 μV at 100°C. It is usual to assume a value of 40 μV for most practical purposes in microclimate work. The traditional technique for using thermocouples involved keeping the reference junction in a thermos flask of ice, which was assumed to be at 0°C, but this is not an ideal arrangement since the greater the difference in temperature between the junctions, the more important the characteristics of the thermocouple become. If the reference junction is kept at about ambient temperature, and its temperature measured by an accurate thermometer, then any error contributed by the thermocouple only applies to the difference in temperature between the two junctions, which will be quite small. If this practice is adopted, it is usually quite safe to assume that the thermocouple is linear: if the difference between the junctions is 5°C then a system that has been calibrated at 20°C will have an error of about 0.2°C at 40°C.

There are combinations of metals that give a greater output than copper – constantan, such as iron – constantan and antimony – bismuth, and some specially produced alloys that have better linearity, but the fact that copper and constantan are so easily available in a wide range of sizes and coverings, and the ease with which they can be soft soldered, make this combination particularly attractive.

It was stated earlier that a thermocouple consists of a pair of junctions between dissimilar metals but in practice a complete thermocouple circuit inevitably includes junctions other than those intended for temperature measurement, such as soldered joints between copper wires, connectors etc. Any junction between dissimilar metals must produce thermoelectric potentials and even a soldered joint in a copper lead must be thought of as a pair of junctions, copper-solder and solder-copper, which will produce potentials that exactly cancel each other, provided there is no temperature gradient at the joint. Joints in thermocouple leads should be made as shown in Fig. 8, rather than being simply butted together, so that the chance of a temperature

Fig. 8. How to make a soldered joint in thermocouple leads.

gradient through the joint is minimised. Connectors can be obtained with pins made of appropriate thermocouple materials, such as copper and constantan, but it is always a good idea to obtain them from the same company that supplies the wire because one is more likely to obtain alloys of exactly the same composition. Similarly, if the same sample of wire can be used throughout the system (this applies particularly to resistance wires such as constantan) then errors due to slight differences in the alloy can be eliminated. If a thin wire is required for the measurement junction and a thicker, flexible lead is more appropriate for the run to the reference junction, a number of strands of the thin wire can be twisted together to make a tougher flexible lead. If three lengths of constantan wire are clamped together at one end in a vice, with the other ends secured in the chuck of a hand drill, the wires can be twisted round each other and a gentle pull will stretch the material just beyond its elastic limit and prevent any tendency to unwind. It is usually convenient to use copper and constantan wires that are covered with solderable synthetic enamel, which as its name suggests melts at the temperature used in soldering.

Thermocouple leads can be joined by connectors that are not of the same material and provided reasonable precautions are taken one usually "gets away with it". A connector that is particularly useful in thermocouple systems where a number of junctions are connected to a measuring box is the 12-way screw connector strip, which is made of plated brass. Provided steps are taken to prevent the sun shining directly onto the connector, and provided the leads are taken right through the connector, as shown in Fig. 9, this practice is usually satisfactory. The use of plugs and sockets made of materials different from the leads is rarely acceptable since the junctions where the spurious potentials can be generated are well separated and even small temperature gradients will cause large errors.

Thermocouples for small-scale microclimatic work can be made from very fine wire (50 swg wire, which is about 0.025 mm, is easily available); they are

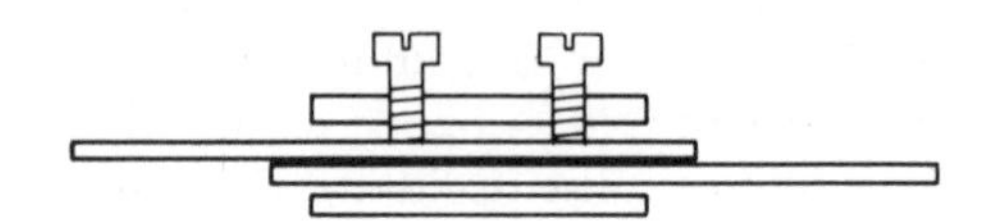

Fig. 9. Joining thermocouple leads in a screw connector strip.

very quick in response and are less likely to give errors due to heat conducted along the leads. Fine thermocouples have a higher resistance but this is no longer a serious problem because modern indicating devices have a high input resistance. Thermocouples that are used in a wet environment should be well insulated to avoid spurious readings caused by electrolytic effects. The voltages produced by thermocouples are very small, and their resistances are not negligible, so that even very small currents fed into the system by the action of an electrolyte on the different metals will cause errors. Methods of encapsulating and insulating temperature probes, including thermocouples, will be described later.

Meters can be purchased to which a single thermocouple can be connected, the reference junction being internal. The reference junction temperature is usually measured by a silicon diode sensor and the meter reading automatically corrected, so that it indicates the measurement thermocouple temperature directly. Such meters are usually made to cover a very wide temperature range and tend to lack the resolution for microclimate work, but progress in this field has been so rapid that one would expect this situation to improve. As explained earlier, it is absolutely essential that the material used in the measurement thermocouple exactly matches the internal junction provided by the makers of the meter.

D. Simple Thermocouple Indicator

Thermocouples give a very small output and in order to drive a moving-coil meter, a considerable amount of amplification is needed. Such amplification can be provided by a simple integrated-circuit operational amplifier, as shown in Fig. 10. The 1 mA meter with its 1k series resistor needs 1 volt for full-scale deflection, but the thermocouple provides an input to the amplifier of 2 mV for a 50°C temperature range; the amplifier therefore requires a voltage amplification of 500. The pre-set potentiometer (S), which should be a multi-turn trimmer type, sets the range covered by the meter, which is 50°C in this design; the circuit can be made to cover other ranges by modification of the circuit values. Changing the value of the 180R resistor connected to the (−) input of the amplifier to 100 R will make the meter cover a range of 25 or 30°C. The 10k, 10-turn helical potentiometer ("Cal") is used to set the meter reading to the temperature indicated on the calibration thermometer, with the input switch (S2) in the "Cal" position. Once this has been done, and with the switch (S2) in the "Use" position, the meter will indicate the temperature of the measurement junction directly. Since the reference junction contributes the same potential in both the "Cal" and "Use" positions of S2, we do not actually need to know its temperature. However, it is important that the temperature of the reference junction does not change quickly, and this

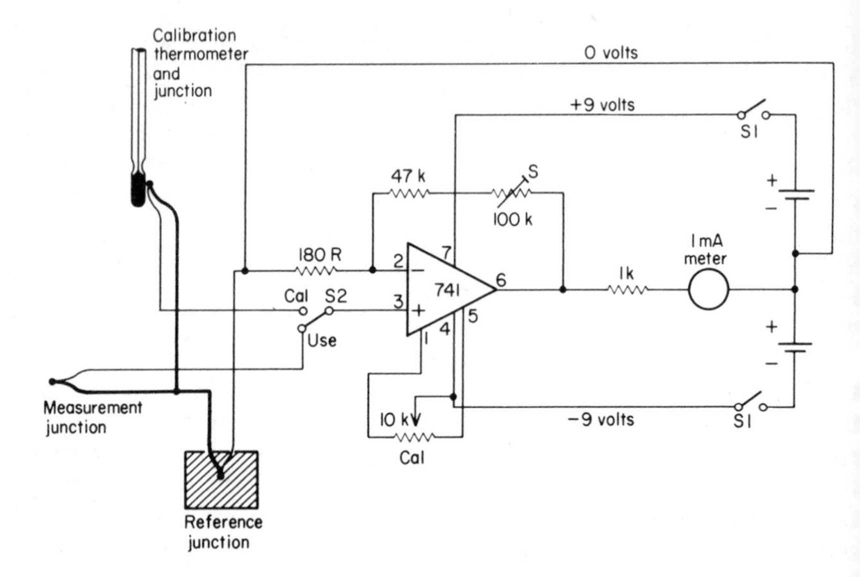

Fig. 10. The simple thermocouple indicator. Thickened lines indicate constantan leads.

can be achieved by inserting it into a block of metal, such as aluminium, and insulating the block is plastic foam.

More than one measurement thermocouple may be used with this indicator simply by connecting all the constantan leads together, and switching the copper leads with a multi-position rotary switch in place of the two-position input switch (S2).

Although Fig. 10 shows a mercury thermometer as the calibration device, there are other possibilities. It may be more convenient to use a bimetallic dial thermometer, which can have the calibration thermocouple fixed inside, close to the sensing element. An electric thermometer, such as the silicon diode meter (Fig. 5), could be used, sharing its meter with the thermocouple indicator with the aid of a change-over switch. Whatever method is used, it is essential that the calibration thermometer and thermocouple are in good thermal contact, and preferably with some insulation around them to reduce the chance of a serious error due to thermal gradients.

E. Temperature Probes and Leads

Whatever type of sensor is used, the electrical connections have to be insulated if it is used in an electrically conducting medium. Resistance thermometers, such as thermistors, will be caused to read inaccurately because of the shunting effect of the resistance placed across the device by leakage through

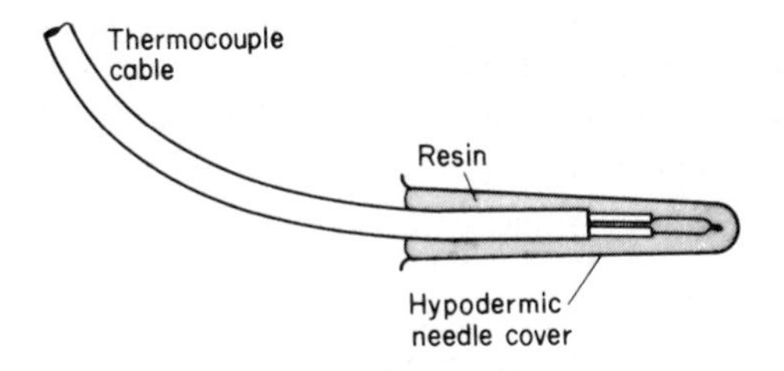

Fig. 11. A technique for making waterproof temperature probes using disposable hypodermic needle covers.

the medium. Thermocouples, which have a low resistance and are unlikely to be greatly affected by the shunt path through the medium, are much more seriously affected by electrolytic effects of water on the various metals in the system, and because of the large amplification that has to be provided, these errors can be very large. When thermocouples are used in water or soil, all parts of the system must be insulated. Special thermocouple wire can be bought, with leads made of copper and constantan (or other combinations) and well insulated by PVC or other suitable materials. If the thermocouple junction is to be buried in the soil, so that the speed of response is unimportant, one of the simplest ways of insulating the junction is to fill one of the protective cases in which disposable hypodermic syringe needles are supplied with epoxy resin and carefully insert the end of the thermocouple cable, with its leads soldered together, into the case. This is shown in Fig. 11, and it is also a useful technique for thermistors or silicon diode sensors.

Some thermistors are supplied with the sensitive element mounted in the end of a glass tube. These are very useful because the leads may be well insulated without seriously affecting the thermal response of the device. In other cases, and generally with silicon diodes, insulation must affect the thermal properties of the sensor. If a thermocouple needs to have a very quick response, or needs to be very small and have low thermal mass, it can be made as shown in Fig. 12. Very fine resin-coated wire of appropriate material is soldered to the ends of the thermocouple cable and the soldered joints well insulated with resin adhesive. Since the fine wire is resin-coated, the only

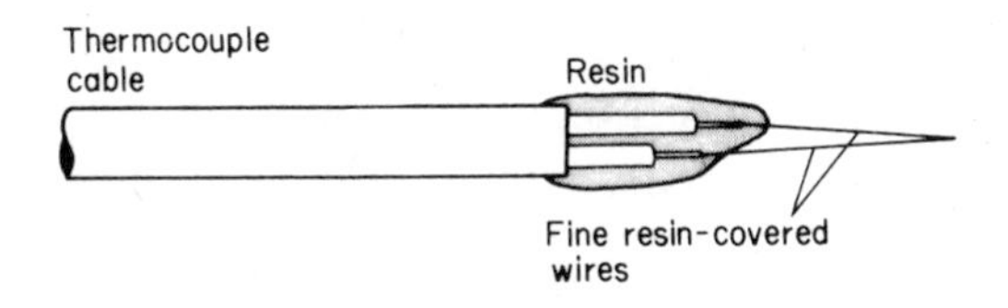

Fig. 12. Very fine thermocouples.

insulation now required is at the junction itself, which can be dipped into thi perspex cement or similar adhesive.

The use of very long leads between the temperature sensor and the indi cator can cause problems due to the resistance of the leads, and the tempera ture coefficient of that resistance. The resistance of thermocouple cables, on lead of which is usually made from a resistance wire like constantan, can b quite large, but with modern indicators this should not cause seriou problems until the cable resistance exceeds 1000 Ω. Cable resistance is no usually a serious problem with thermistor systems, provided thermistors o reasonably high value are used. Twin 16/0.2 mm flat mains cable has a resis tance of about 4 ohms per 100 metres: a 5000 ohm thermistor would not hav its performance seriously affected by 100 metres of such cable. Metal resis tance thermometers generally have a low resistance, and special precaution have to be taken, such as the provision of separate leads for feeding curren through the element and measuring the voltage.

IV. Testing and Calibration of Temperature Sensors

Temperature sensors should always be calibrated in a liquid. Water is the obvious choice provided the sensor leads are well insulated, but light oil should be used if the leads are uninsulated and intended only for use in air. The liquid should be continuously stirred while the measurements are being made, preferably with a magnetic stirrer.

If the sensor is for use in air, a check should be made to see that self-heating does not seriously affect the reading in still air. This test has to be carried out in an environment that contains both air and water at the same temperature. This can be achieved by partly filling a bottle with water and suspending the sensor just above the water surface with its lead passing through a well-fitting hole in the stopper. The bottle should be left for at least an hour with the sensor connected to its indicator and switched on, preferably in a constant temperature cabinet or at least in a room where the temperature changes only slowly. When a steady state has been achieved, the sensor should be lowered into the water, and any change in reading noted. A good temperature meter should not change its reading by more than 0.1°C.

Defects in mercury thermometers are easily seen but other types of thermometer, particularly electrical types, can give errors that are not easy to detect and it is therefore strongly recommended that all temperature meters, whether bought or made, should be regularly checked against a good quality mercury-in-glass thermometer.

2
Radiation

I. The Measurement of Radiation

The wavelength of maximum radiation of a black body radiator is inversely related to its absolute temperature (Wien's displacement law). The sun has a surface temperature of about 6000 degrees Kelvin (K) and a wavelength of maximum emission of 0.48 μm. Most of the energy in the sun's radiation lies within the range 0.2 to 4 μm and is called "short-wave" radiation. The earth has a surface temperature of about 300 K, and its wavelength of maximum emission is about 10 μm. Most of the energy in terrestrial radiation lies in the range 3 to 80 μm and is called "long-wave" radiation. Figure 13 illustrates the terms used to describe parts of the solar and terrestrial radiation spectrum, and the range of wavelengths to which radiation instruments are sensitive.

The part of the spectrum to which our eyes are sensitive, which we call light, extends from 0.4 to 0.7 μm. Wavelengths shorter than 0.4 μm are called "ultra-violet" and wavelengths longer than 0.7 μm are called "infra-red". The part of the infra-red spectrum with wavelengths less than 3 μm is called "near infra-red" to distinguish it from the long-wave part, or "far infra-red".

A solarimeter (or pyranometer) measures total solar radiation, both direct and diffuse, received by a horizontal surface. Because the amount of radiation collected by the horizontal surface will depend on the angle of the sun, these instruments are usually covered by a hemispherical glass dome. A pyrheliometer also measures short-wave radiation, but at an angle normal to the sun and with diffuse sky radiation excluded. Light meters (photometers) are sensitive to the visible parts of the solar spectrum, and exclude ultra-violet and infra-red.

Pyrradiometers measure the total short-wave and long-wave radiation and

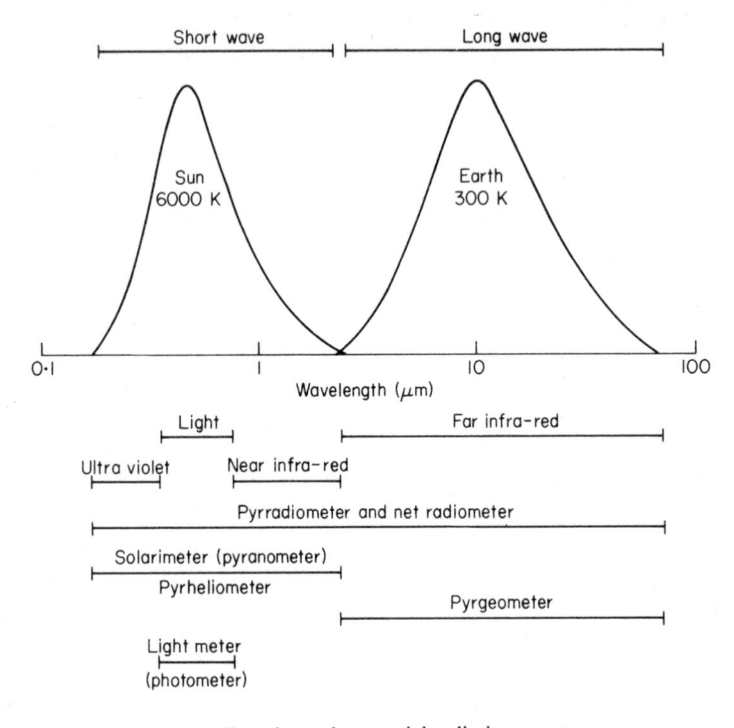

Fig. 13. The solar and terrestrial radiation spectrum.

pyrgeometers measure only long-wave radiation. Net radiometers measure the difference between the total short-wave and long-wave radiation received by a surface and the total short-wave and long-wave radiation reflected or radiated by it.

Some radiation instruments are rarely used by ecologists. Pyrheliometers are normally used only for calibration purposes. Pyrradiometers are rarely used because the same information can be obtained by fitting a net radiometer with a metal dome on its lower surface. The radiation emitted by the dome can be calculated from its temperature, and the total short-wave and long-wave radiation is equal to the sum of this value and the net radiation. Pyrgeometers are rarely used since the same information can be obtained by subtracting the reading of a solarimeter from that obtained from a net radiometer fitted with a dome on its lower surface. The three instruments that are generally of interest to ecologists are therefore the solarimeter, the light meter and the net radiometer.

A. Solarimeters

Conventional solarimeters (or pyranometers) generally use a number of hermocouples in series (collectively, a thermopile) to measure the difference in temperature between black and white areas on a flat plate. The plate is covered by a hemispherical glass dome which transmits only the short-wave spectrum (0.2 to 4.5 μm for quartz glass). A typical instrument of this type is he Eppley pyranometer, which is a beautiful and accurate instrument, and is often used as a secondary standard against which other radiation instruments are calibrated.

Silicon solar cells may also be used for measuring solar radiation. They do not have such a flat spectral response as conventional solarimeters but the output is very much larger (at least 10 000 times). A special glass dome is not essential since the cells are not sensitive to longwave radiation. Simple radiometers can be made using solar cells.

Solarimeters have been made in tubular form for insertion, horizontally, under a crop canopy. These are inherently less accurate than instruments with a hemispherical dome, but can be of great use in estimating the average radiation below a canopy relative to that above.

Solarimeters measure the total shortwave radiation, but if they are fitted with a device to cast a shadow on the sensitive area, so that the direct beam is eliminated, they will indicate only the diffuse (sky) radiation. Such a device is sometimes made in the form of a narrow band called an "occulting ring". The energy of the direct beam may be calculated by subtracting the diffuse reading from the total radiation.

Solarimeters can also be used to measure the short-wave radiation reflected from a surface. The sensor is inverted, fitted with a shield to eliminate diffuse sky radiation, and mounted high enough over the surface so that the shadow it casts is a very small part of the surface area being investigated. In such a

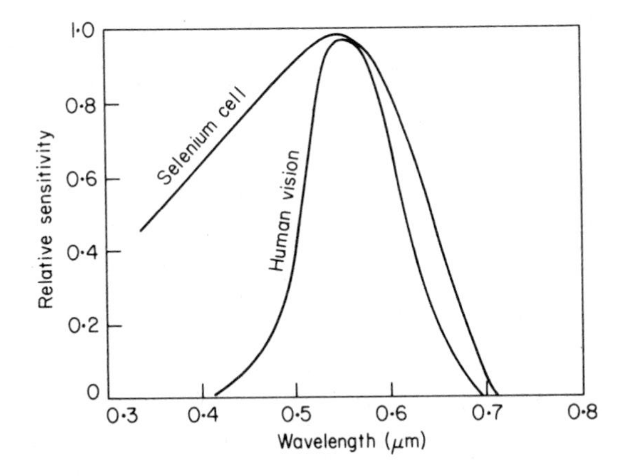

Fig. 14. Approximate spectral response of the human eye and a selenium photocell.

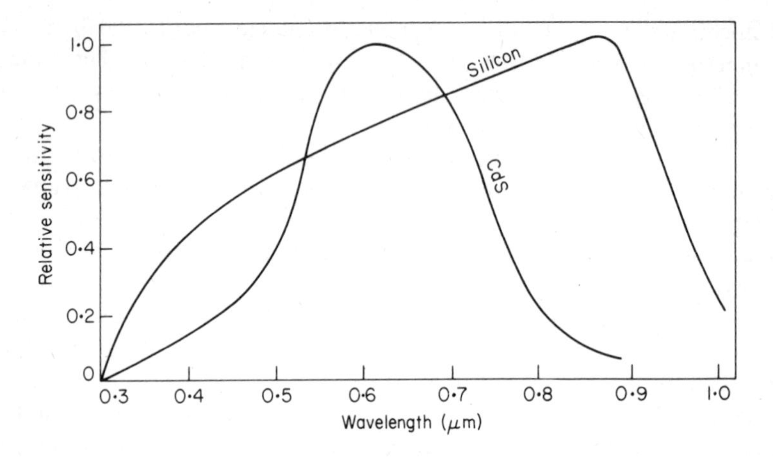

Fig. 15. Approximate spectral response of a silicon photo-diode and a cadmium sulphide photoconductive cell.

configuration, a solarimeter is often called an "albedometer". (The reflection coefficient of a surface is sometimes called the "albedo".)

B. Light Meters

A light meter is a solarimeter with the spectral response limited to the visible range (0.4 to 0.7 *μm*) although the term is often loosely applied to such devices as the "ultra-violet light meter". Light measurements can be made by fitting a solarimeter with appropriate filters (Szeicz, 1966) but it is usually more convenient to use some kind of photocell. The selenium photocell has a spectral response similar to the human eye (Fig. 14) but it does not respond linearly, it changes its calibration with time, and it is insensitive to low light intensities. The cadmium sulphide photoconductive cell covers a range of about 0.3 to 0.9 *μm* (Fig. 15) and some of the better-quality photographic light meters incorporate a filter to limit the spectral response to the visible range. A correcting filter can also be used with silicon photo-diodes (McPherson, 1969).

C. Net Radiometers

The earth's surface receives short-wave radiation from the sun. Some of this is reflected away, and the remainder is absorbed and raises the temperature. The earth's surface also emits long-wave radiation. A net radiometer is mounted above a surface being studied, at a sufficient height so that the shadow cast by the radiometer covers only a small part of the area "seen" by the lower window of the instrument. The upper window is illuminated by the incoming radiation and the output of the radiometer is proportional to the net radiation (difference between total input and output) of the surface. If the net radiation is positive, then the surface is gaining heat and its temperature will rise. If the net radiation is negative, which often happens at night and in the shade on cloudless days even in summer, the surface is losing heat and the temperature will fall.

Net radiometers are double-sided, unlike all the other radiation instruments, and consist of a black metal plate fitted with temperature sensors to measure the difference in temperature between the upper and lower surfaces. In some designs, the plate is evenly ventilated and in others both sides are covered by a thin plastic shield, usually polythene. Net radiometers are usually circular, but special tubular versions have been made for average measurements in crop canopies. These are inserted horizontally in the same way as tube solarimeters, and are best used for relative rather than absolute measurements.

II. Major Sources of Error in Radiation Measurement

The most important sources of error in radiation measurement are concerned with the spectral sensitivity of sensors. A light meter should be sensitive to all wavelengths in the visible range, and not sensitive to wavelengths outside this range. Many light meters have an uneven response in the visible range and respond to wavelengths outside it. However, a meter that is calibrated in, for example, sunlight will give an accurate reading in light of similar spectral distribution, even if the spectral sensitivity of the sensor is somewhat uneven. It is when light of different spectral distribution (or light that has been modified by selective reflection or transmission) is being measured that errors will occur. In a similar way, we often do not know the range of wavelengths to which living organisms respond and when we look for correlations between radiation measurements and the behaviour of living organisms, the question of spectral sensitivity should always be considered. Even if we get the radiation measurements right, we may not be measuring the same thing as the animal!

Solarimeters can have a similar problem with spectral sensitivity. Conventional solarimeters (thermopile type) should be fitted with special glass domes if they are to respond over the appropriate range for short-wave solar radiation (0.2 to 4μm) and exclude long-wave radiation (4 to 80 μm). Silicon solar cell solarimeters, which are very simple and sensitive instruments, do not have an even response over the range required but include about 90% of the sun's radiation, with a peak at about 0.9 μm (Fig. 15). Provided the spectral distribution of the radiation being measured is similar to that emitted by the sun, such instruments will be quite satisfactory, but errors will result if the distribution has been modified, for example by passing through a leaf. Since leaves absorb mainly the shorter wavelengths, and silicon solar cells have their maximum response at the longer wavelengths, they will give too high a reading of radiation transmitted through a leaf. The error is not large (about 5% in some cases) and such measurements may be useful for comparative purposes, but a thermopile solarimeter is a better instrument for this type of work.

Most types of solarimeter, light meter or net radiometer measure the radiation incident on a horizontal surface from a complete hemisphere and should give an output that is proportional to the cosine of the angle of the radiation relative to a line normal to the surface. An instrument that behaves in this way is said to have a "good cosine response". Instruments that do not have a good cosine response, often due to reflection of radiation from the surface of the sensor, usually have their errors confined to angles near the horizontal (approaching 90 degrees relative to normal incidence) from which direction there will be very little radiation in many cases. However, measurements at dawn or dusk in open situations could be affected by any serious departure from a cosine response.

There may be occasions where the measurement of radiation on a surface normal to the direction of the major source (usually the sun) is more meaningful than that on a horizontal surface. If such readings are taken, it should always be stated that they represent radiation "at normal incidence".

III. Simple Equipment for Radiation Measurement

A. Light Meters

1. Photographic exposure meters. Photographic exposure meters are made for a mass market and therefore cost very much less than scientific instruments. Some are very sensitive and can measure down to the level of moonlight, and in order to expose colour film correctly, they need to have a reasonable degree of accuracy. There are two types of exposure meter; those which are pointed at the light source (incident light reading) and measure illumination, and those which are pointed at the subject being photographed (reflected light reading) and measure luminance (surface brightness). Photographers need to know the brightness of the subject, as measured directly by the reflected light meter, but incident light meters measure illumination and make an allowance for the reflectance of the subject (usually 18% reflectance is assumed) to get an estimate of luminance. The readings of photographic exposure meters are usually expressed in terms of exposure time and lens aperture, for a pre-set value of film speed. Some meters also have an "exposure value" (EV) scale, again for a pre-set value of film speed. The exposure value is related to exposure time and lens aperture by the equation:

$$EV = 3.322 \log_{10}(n^2/t)$$

where n is the aperture of the lens (f number) and t is exposure in seconds.

Ecologists are generally more interested in illumination than luminance, and an incident light meter (or a reflected light meter with an incident light adaptor) is therefore the more useful type. However, reflected light meters (or cameras with built-in light meters, which are usually of the reflected light type) can be used to estimate incident light by measuring the light reflected from a surface with a reflectance of 18%. Good quality white writing paper has a reflectance of about 80%. If a sheet of white paper is illuminated with a constant light, and the exposure time and film speed controls of a reflected light meter (or camera) are adjusted so that the aperture indicated is $f/11$, then a grey surface of 18% reflectance would give a reading just under $f/5.6$ (actually about one-fifth of the way from $f/5.6$ to $f/4$). Using these figures, a piece of card with 18% reflectance may be found, and used as a reference for estimates of incident light.

The readings of either type of photographic exposure meter can be expressed in light units; reflected light readings can be expressed in luminance units, such as candelas per square metre (cd/m^2), and incident light readings can be expressed in illumination units, such as the lux. Tables 1 and 2 give these conversions from photographic readings to light units, including SI, CGS and imperial systems. There are some approximations involved in the photographic readings, the most serious being the lens aperture calibration (f numbers) which are not in an exact series, but the tables and the formulae given below are accurate to about 5%.

(i) Luminance from reflected light readings (Table 1)

$$L = 0.0014 \times \text{film speed} \times \frac{n^2}{t}$$

where L is the luminance in candelas per square metre (sometimes called the "nit"), film speed is in ASA units, n is the lens aperture (f number) and t is the exposure time in seconds.

(ii) Illumination from incident light readings (Table 2)

$$I = 0.0244 \times \text{film speed} \times \frac{n^2}{t}$$

where I is the illumination in lux (lumens per square metre), film

Table 1

Luminance from reflected light readings. The preferred unit is the candela per square metre. Film speed set to 100 ASA[a]

Exposure time (seconds)									
f/2	*f*/2·8	*f*/4	*f*/5·6	*f*/8	*f*/11	cd/m^2 (nit)	cd/ft^2	foot – lambert	Exposure value
16	32	64	–	–	–	0·035	0·0033	0·01	–2
8	16	32	64	–	–	0·07	0·0065	0·02	–1
4	8	16	32	64	–	0·14	0·013	0·041	0
2	4	8	16	32	64	0·28	0·026	0·082	1
1	2	4	8	16	32	0·56	0·052	0·16	2
1/2	1	2	4	8	16	1·1	0·10	0·33	3
1/4	1/2	1	2	4	8	2·2	0·21	0·65	4
1/8	1/4	1/2	1	2	4	4·5	0·42	1·3	5
1/16	1/8	1/4	1/2	1	2	9·0	0·83	2·6	6
1/32	1/16	1/8	1/4	1/2	1	18	1·7	5·2	7
1/64	1/32	1/16	1/8	1/4	1/2	36	3·3	10	8
1/125	1/64	1/32	1/16	1/8	1/4	70	6·5	20	9
1/250	1/125	1/64	1/32	1/16	1/18	140	13	41	10
1/500	1/250	1/125	1/64	1/32	1/16	280	26	82	11
1/1000	1/500	1/250	1/125	1/64	1/32	560	52	160	12
1/2000	1/1000	1/500	1/250	1/125	1/64	1100	100	320	13
–	1/2000	1/1000	1/500	1/250	1/125	2200	200	640	14
–	–	1/2000	1/1000	1/5000	1/250	4500	420	1300	15
–	–	–	1/2000	1/1000	1/500	8500	790	2500	16
–	–	–	–	1/2000	1/1000	17000	1600	4900	17
–	–	–	–	–	1/2000	36000	3300	10000	18

[a]For film speed settings other than 100 ASA, multiply the result by 100/film speed.

Table 2

Illumination from incident light readings. The preferred unit is the lux. Film speed set to 100 ASA[a]

Exposure time (seconds)									
$f/2$	$f/2{\cdot}8$	$f/4$	$f/5{\cdot}6$	$f/8$	$f/11$	Lux (lm/m^2)	milliphot ($10^{-3}\cdot lm/cm^2$)	Foot-candle (lm/ft^2)	Exposure value
16	32	64	–	–	–	0·61	0·061	0·057	−2
8	16	32	64	–	–	1·2	0·12	0·11	−1
4	8	16	32	64	–	2·4	0·24	0·23	0
2	4	8	16	32	64	4·9	0·49	0·45	1
1	2	4	8	16	32	9·8	0·98	0·91	2
1/2	1	2	4	8	16	20	2·0	1·8	3
1/4	1/2	1	2	4	8	39	3·9	3·6	4
1/8	1/4	1/2	1	2	4	78	7·8	7·3	5
1/16	1/8	1/4	1/2	1	2	160	16	15	6
1/32	1/16	1/8	1/4	1/2	1	310	31	29	7
1/64	1/32	1/16	1/8	1/4	1/2	620	62	58	8
1/125	1/64	1/32	1/16	1/8	1/4	1200	120	110	9
1/250	1/125	1/64	1/32	1/16	1/8	2400	240	230	10
1/500	1/250	1/125	1/64	1/32	1/16	4900	490	450	11
1/1000	1/500	1/250	1/125	1/64	1/32	9800	980	910	12
1/2000	1/1000	1/500	1/250	1/125	1/64	20000	2000	1800	13
–	1/2000	1/1000	1/500	1/250	1/125	39000	3900	3600	14
–	–	1/2000	1/1000	1/500	1/250	78000	7800	7300	15
–	–	–	1/2000	1/1000	1/500	160000	16000	15000	16
–	–	–	–	1/2000	1/1000	300000	30000	28000	17
–	–	–	–	–	1/2000	590000	59000	55000	18

[a]For film speed settings other than 100 ASA, multiply the result by 100/film speed.

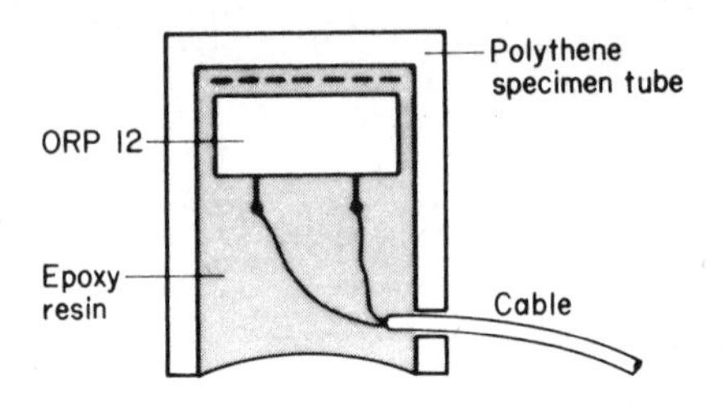

Fig. 16. Construction of the light meter. The filter is shown in broken lines.

speed is in ASA units, n is the lens aperture (f number) and t is th exposure time in seconds.

2. *CdS cell light meter.* Cadmium sulphide photoconductive cells ar devices whose resistance is determined by the level of incident radiation. The have a spectral response that includes the visible range and extends into th near infra-red, but filters can be used to limit it to the visible range. Th simple light meter described below will measure over a range of 10 to 100 00 lux and the head which contains the photocell is designed so that it can b used under water, as well as in air.

The construction of the head is shown in Fig. 16. The end part of a poly thene specimen tube, large enough so that it just fits over the ORP 12 cell, i used as a diffuser. The sides of the cell are painted black so that light can onl reach the sensitive area through its end window, and hence through the filte Special filters can be expensive, but the Kodak Wratten 38, which is availab in gelatin form at low cost, is sufficiently close to the desired characteristics To assemble the head, the end part of the specimen tube is inverted and a dis

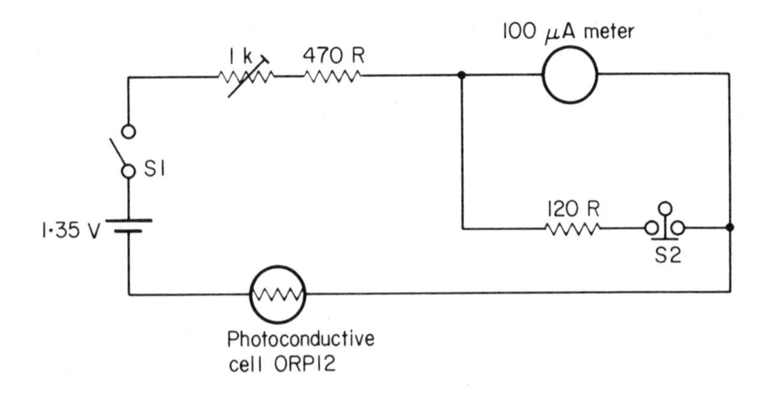

Fig. 17. Circuit diagram of the light meter.

f gelatin filter laid in the bottom. The ORP 12 cell is laid on the filter and ired to the cable. The back of the assembly is then filled with epoxy resin to old the components in place and make the head waterproof.

The circuit of the light meter is shown in Fig. 17. Power is taken from a .35 V mercury battery (as used to power the light meters of many cameras) nd the pre-set trimmer resistor adjusted so that the meter reads full-scale hen the leads to the cell are connected together. No further adjustment iould be necessary, as mercury batteries have a very stable voltage, up to the oint of failure. The push switch (S2) is wired so that its contacts are normally osed, and open with pressure on the button when the low-light range is elected. The overall performance of the prototype meter is shown in Fig. 18.

B. Solarimeters

1. Silicon solar cell solarimeter. If a silicon solar cell is connected to a low esistance measuring circuit, the current produced is proportional to the icident radiation over a wide range. This fact, combined with the broad (if omewhat uneven) spectral response enables us to use the solar cell as the ensing element in a very simple solarimeter. In the circuit of Fig. 19, the ush switch (S) is connected so that it is normally closed, resulting in a resis-nce of 6.8 Ω being connected across the solar cell. The voltage developed cross this resistance by the current from the cell is measured by the micro-

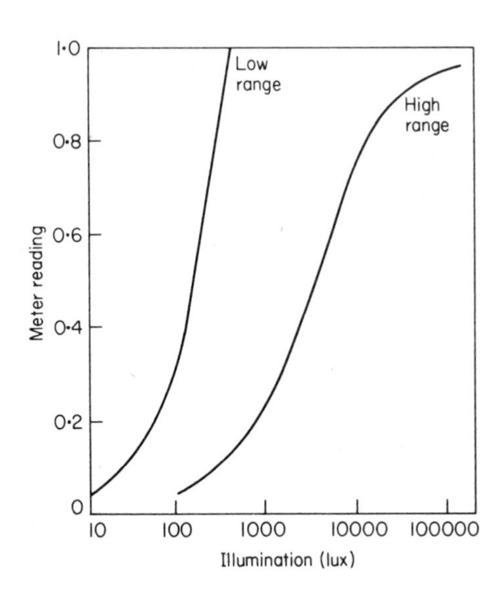

Fig. 18. Characteristics of the light meter.

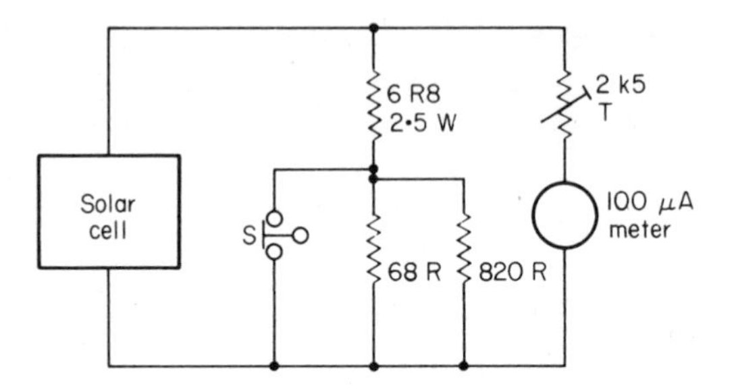

Fig. 19. Circuit diagram of the solar cell solarimeter.

ammeter and its sensitivity can be set by adjusting the pre-set trimmer resisto (T). Pressing the switch (S) increases the total load resistance on the solar cel by a factor of 10, resulting in a 10-fold increase in sensitivity.

The construction of the head in shown in Fig. 20. The cell is mounted on a aluminium plate in which is cut a rectangular aperture corresponding to th sensitive area of the cell. A plastic diffuser is cemented to the upper side of th plate. The prototype used a piece of white plastic from the cap of an aeroso can, which had a thickness of about 1 mm, as a diffuser. The upper surfac of the diffuser should be roughened with fine abrasive paper and the part of the surface outside the aperture in the aluminium plate, and the edges o the diffuser, should be painted black. This procedure will ensure that th solarimeter has a good cosine response.

2. *Thermopile solarimeter*. The sensor of the thermopile solarimete consists of a thin disk of insulating material, and a large number of thermo couples, all connected in series, measuring the difference in temperatur between black and white areas painted on the disk. The arrangement of th thermocouples is shown in Fig. 21. The disk should be made from thi insulating material, and 0.5 mm Tufnol or Paxolin is ideal, but thin car which has been stiffened and made waterproof by the application of severa coats of varnish will be satisfactory. The holes (1 mm) should be drilled in th disk and then a length of 44 swg uncoated constantan wire can be lace through the holes, starting at the outside (top in Fig. 21) and finishing in th

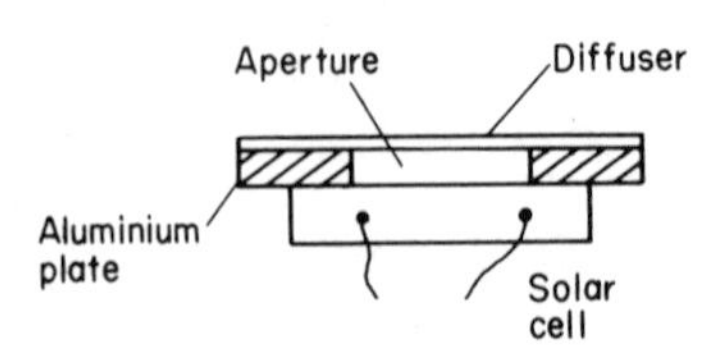

Fig. 20. Mechanical arrangement of the solar cell and diffuser.

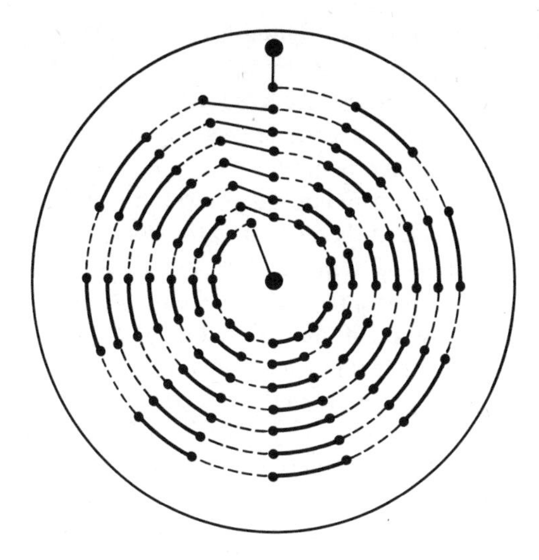

Fig. 21. The solarimeter disk. Broken lines indicate constantan wires, solid lines indicate copper wires.

ntre, passing alternately above and below the disk. The constantan wires ould then be cemented to the underside of the disk, and the wires on the per side cut and replaced by copper wires, forming pairs of thermocouple ıctions. The copper wire should preferably be uncoated, and suitable

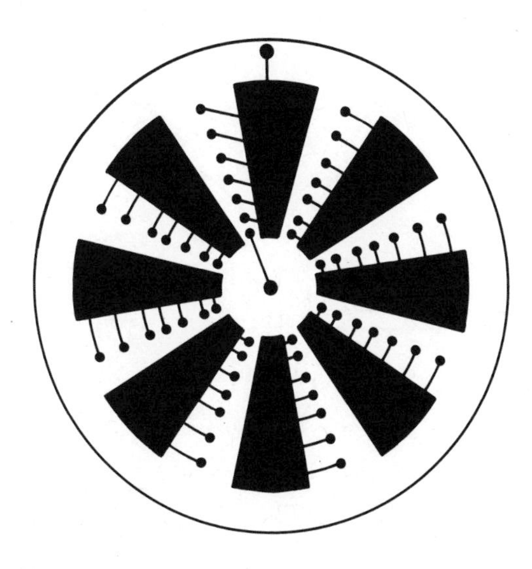

Fig. 22. Location of the black segments on the solarimeter disk.

material can be found in the screening layer in many co-axial or screen cables. It should have a thickness similar to that of the constantan wire (swg). The junctions should be soldered with great care, and unless t eyesight of the constructor is unusually good at close distances, a ster dissecting microscope should be used.

When the assembly of thermocouples (thermopile) is complete, the copp wires should be cemented to the disk, and the black and white areas can painted, as shown in Fig. 22. Optical black paint is ideal for the black area but blackboard paint is also satisfactory. The paint used for the white are should also have a matt finish, and some typewriter correcting fluids (e Tipp-Ex) are excellent for this purpose.

The completed disk is mounted in the solarimeter housing by cementing to the tips of three screws with epoxy adhesive, as shown in Figs 23 and 2 The back of the solarimeter is fitted with an aluminium plate and the fro with a flat glass plate. The use of a flat glass will cause a departure fro a cosine response at low angles of radiation and, if this is likely to be importance, a dome should be fitted.

Thermopile solarimeters vary greatly in size, from about 15 to 100 mm diameter. The larger ones, which have the largest number of thermocoup junctions, are much more sensitive than the smaller ones, which may ha as few as eight junctions. Having decided on the approximate size fo solarimeter, the most difficult problem is usually the acquisition of a suital dome or glass plate. If a quartz dome can be found, then this will be ideal a the other components should be made of an appropriate size for the dom A dome of ordinary glass, which will cause some loss of sensitivity at t

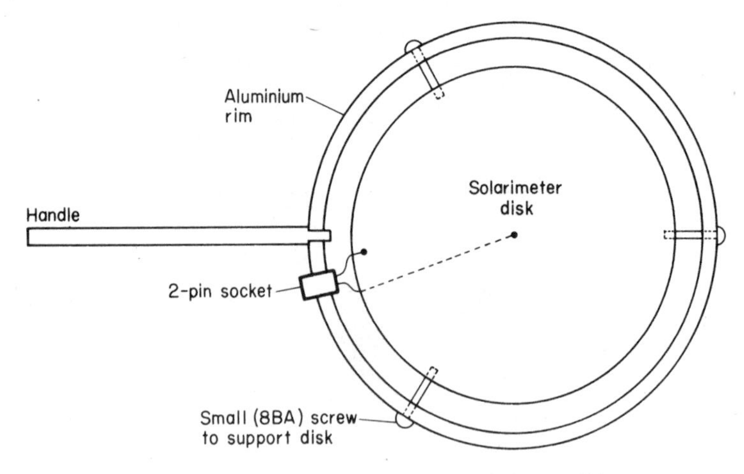

Fig. 23. Method of mounting of the solarimeter disk.

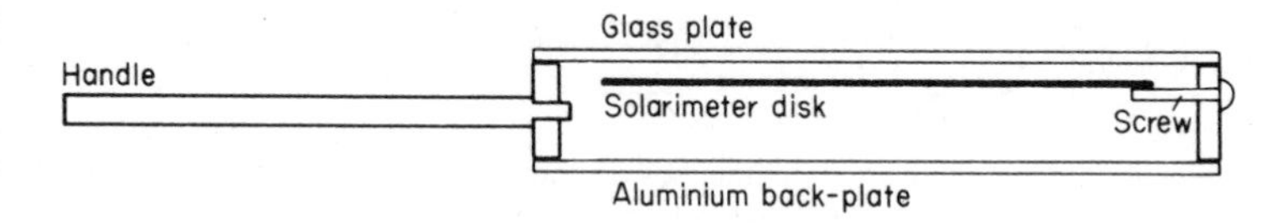

Fig. 24. Section of the complete solarimeter.

xtreme ends of the spectrum, can be made by cutting the end from a large yrex centrifuge tube. The number of junctions used under such a dome vould be less than is illustrated in Fig. 23, and the sensitivity would be corres-ondingly reduced, but a very satisfactory solarimeter can be made with such dome. The disk shown in Fig. 23 was made to fit the circular glass from a ycle front lamp, and the complete solarimeter was 70 mm in diameter. It had sensitivity of 5.2 μV per W m^{-2}.

The spectral response of thermopile solarimeters may be modified by the se of filters (Szeicz, 1966), so that the energy in different parts of the pectrum may be measured.

C. *Net Radiometer*

he net radiometer consists of a rather thicker disk than that used in the olarimeter, with a large number of thermocouples connected so that they neasure the difference in temperature between the two sides of the disk, both f which are painted black. The disk is made from 3 mm Bakelite (any nsulating material which would stand the temperature of soldering would be atisfactory) and a pattern of holes (1 mm diameter) is drilled as evenly as ossible, as shown in Fig. 25. The disk illustrated is a large one, 65 mm in

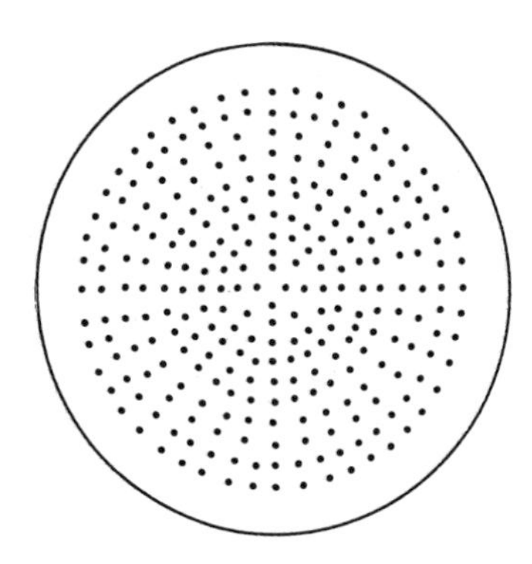

Fig. 25. Arrangement of holes in the net radiometer disk.

Fig. 26. Section of part of the net radiometer disk showing method of wiring. Broken lines indicate constantan wires, solid lines indicate copper wires.

diameter, and a smaller radiometer would have less holes, with a corresponding reduction in sensitivity. The wire used is the same as in the solarimeter described earlier (44 swg uncoated) and the thermocouples are wired as shown in Fig. 26. The completed disk is coated with adhesive on both side to secure the thermocouples to the disk, and painted with matt black paint. If it is available, optical black is ideal, but as with the solarimeter, blackboard paint can be used.

The disk is mounted in a similar way to the solarimeter disk, but to preserve symmetry, the supporting screws fit into shallow holes drilled into the edge, and the disk is mounted equidistant from the upper and lower faces of the aluminium rim (Fig. 27). The glass and aluminium plates of the solarimeter are replaced by thin plastic membranes; glass cannot be used since it does not transmit long-wave radiation. Since thin plastic membranes deteriorate quite rapidly and have to be replaced, there is a good deal to be said for using clear self-adhesive tape as a membrane in small net radiometers. This can be obtained in widths up to 40 mm, and this would be a good size for a net radiometer.

The net radiometer illustrated in Figs 25 to 27 had a sensitivity of 7.8 μV per W m^{-2}.

D. Millivoltmeter for Use with Thermopile Instruments

The circuit of a simple millivoltmeter is given in Fig. 28. It has three ranges (1, 3, 10 mV) which are suitable for use with the thermopile instruments described in this chapter. Additional ranges can be fitted by exchanging the three-position range switch for one with more positions and providing additional range resistors. The values of these can be calculated as follows:

$$R = 100 \left(\frac{1000}{v} - 1\right)$$

Fig. 27. Section through the net radiometer.

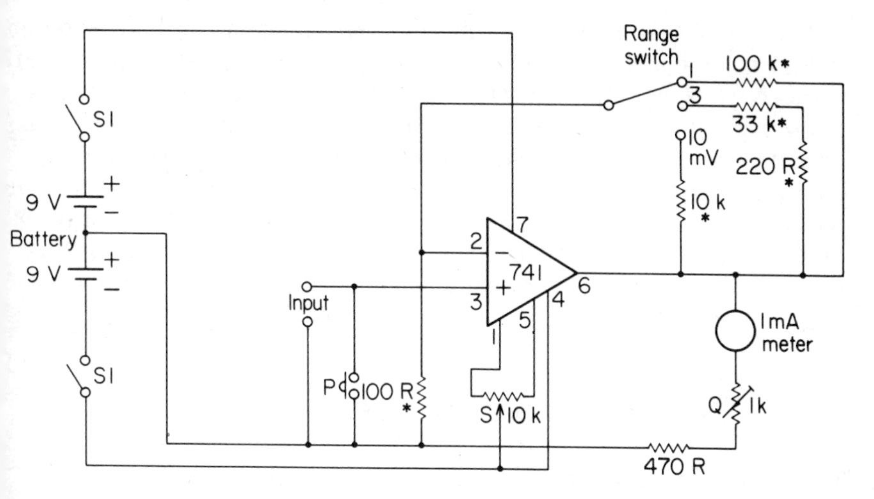

Fig. 28. Circuit diagram of a millivoltmeter for use with the solarimeter and net radiometer.

where R is the value of the range resistor in ohms and v is the input sensitivity in millivolts.

The 741 operational amplifier used in this circuit has a temperature coefficient of 5 μV/°C, which would result in a zero error if uncorrected, and the potentiometer (S) and the input short-circuiting push-button (P) are provided so that the zero can be checked and corrected. The potentiometer (S) is a 10-turn helical type, mounted on the front panel, but provided with a lock since it will not require frequent adjustment.

If the millivoltmeter is to be used with a net radiometer, whose output may be negative, some provision must be made to indicate negative voltages. A reversing switch (Fig. 29) may be fitted in the leads to the meter, or the meter may be scaled so that it has an elevated zero (say -0.2 to 1.0 mV with zero one-sixth of the way up the scale).

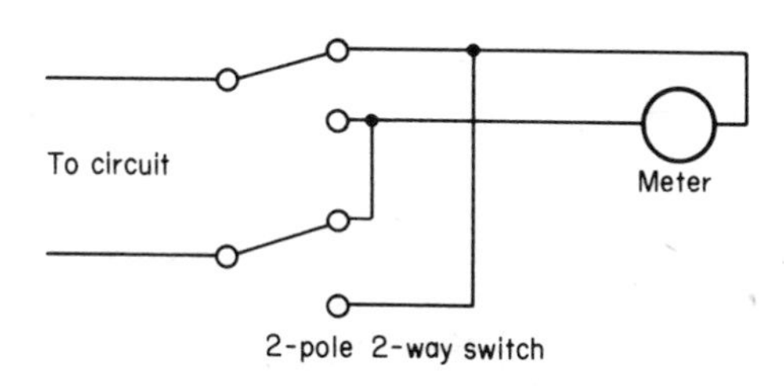

Fig. 29. Reversing switch for the millivoltmeter.

The overall sensitivity of the millivoltmeter may be set by adjusting the pre-set potentiometer (Q). The relationship between the three (or more) ranges is set by the values of the resistors marked on Fig. 28 with an asterisk. If these are of the close tolerance type (2% or 1% metal oxide), individual adjustment of ranges will be unnecessary. An adequate voltage standard can be made by connecting resistor of 1.2k, 150 Ω and 1Ω (close tolerance) in series across a 1.35 V mercury battery. A current of 1 mA will flow through the resistors, producing a voltage drop of 1 mV across the 1 Ω resistor. Other voltages can be obtained by changing the 1 Ω resistor for a different value, the voltage produced being equal to the value of the resistor in ohms.

IV. Testing and Calibration of Radiation Instruments

The best advice that can be given to anyone who wishes to calibrate a radiation instrument is that they should compare it with a professional instrument which has been calibrated against a standard. Comparisons should always be made using a radiation source appropriate to the instrument (e.g. calibrate a solarimeter on sunlight). This minimises errors due to uneven spectral response. In many cases, however, a radiation instrument which is not accurately calibrated will still be of considerable use. It can be used for comparative measurements such as reflectivity (albedo) and transmission and will enable the ecologist to plot the solar input throughout the day in relative terms. An uncalibrated net radiometer will still tell the user whether heat is being gained or lost by the system, and temperature measurements will be more easily interpreted with knowledge of the changes in net radiation.

For the purposes of testing, rather than calibration, of radiation instruments, some "rules of thumb" can be suggested, and these are listed below:

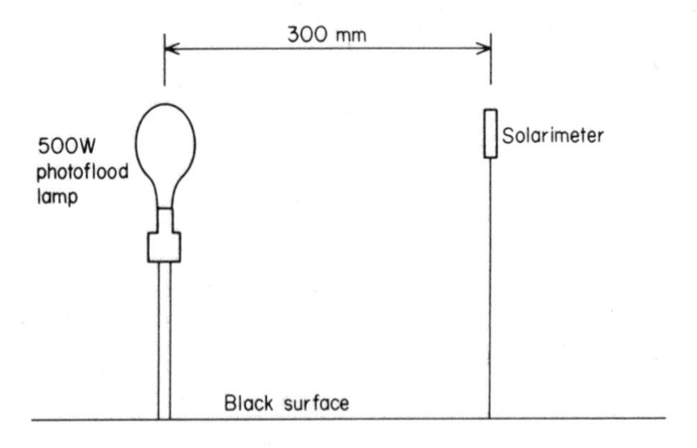

Fig. 30. Set-up for testing a solarimeter.

(i) Sunlight on a clear day in summer in Cambridge measures about 90 000 lux, 800 W/m^2; and a piece of white writing paper illuminated by the sun (at normal incidence) has a surface brightness of about 23 000 cd/m^2.

(ii) The surface brightness of the coating of a 40 W warm white fluorescent tube is about 6 700 cd/m^2 (Weast, 1979); that of an 80 W, 5 ft tube (warm white) measures 6500 cd/m^2.

(iii) A solarimeter placed 300 mm from the centre line of a 500 W photoflood lamp indicates 220 W/m^2 (Fig. 30).

When testing a solarimeter, its sensitivity to long-wave radiation should be checked. As a rough guide, a 60 W soldering iron held 3 mm in front of the sensor should not cause a deflection on the meter, when set to a sensitivity suitable for measuring sunlight. If a thermopile solarimeter does show a deflection, it may be caused by an inadequate thermal contact between the dome (or glass) and the body of the instrument.

3
Humidity

I. The Measurement of Humidity

The measurement of atmospheric water content is very important to ecologists. Plants and animals are affected by changes in humidity and they can themselves cause local modifications to the humidity: much the same can be said of humidity sensors. Some sensors remove water from the air, some evaporate water into it, some can do both; and when humidity is being measured on a very small scale, these modifications to the microclimate caused by the measurement process are often a larger source of error than any shortcomings in the instruments themselves. Some of the more traditional methods, such as the psychrometer and dew-point, have the considerable advantage that provided the instruments are properly constructed, they enable us to make a very accurate estimate of humidity from a pair of temperature readings. Since temperature standards are very much easier to maintain than humidity standards, which will be discussed later in this chapter, these methods of humidity measurement are often used as standards against which other types of instrument are calibrated. With possibly only one exception, methods of humidity measurement that do not involve the measurement of temperature have the disadvantage that instruments have to be calibrated individually by comparison with an instrument such as a psychrometer or against a set of standard humidities. In particular, some of the recent high-technology instruments, which are in most respects more convenient to use than the earlier methods and cause minimal change to the local environment, do need regular calibration. The provision of a meter scale or digital display directly calibrated in relative humidity is no guarantee of accuracy.

A. *The Psychrometer*

The psychrometer consists of two thermometers, one of which measures air temperature while the other is covered by a water-soaked wick and indicates a temperature lowered by evaporative cooling. The difference in temperature between the two thermometers is called the wet bulb depression and is related to the degree of saturation of the air. Water is evaporated into the air from the wet bulb, with a consequent lowering of its temperature, until equilibrium is reached, and provided the rate of aspiration exceeds 3 metres per second (m/s), its value does not affect the wet bulb depression. Psychrometers in which the rate of aspiration is determined by mechanical means are usually called "aspirated psychrometers", and those in which naturally occurring air currents are relied on for ventilation, as in the psychrometers used in the Stevenson screen, are called "non-aspirated".

With non-aspirated psychrometers, such as the one shown in Fig. 31, it is usual to assume that the rate of air movement past the thermometers is 1 m/s, at which rate the wet bulb depression reaches 83% of the value it would achieve when aspirated at a rate in excess of 3 m/s. In very still air, non-aspirated

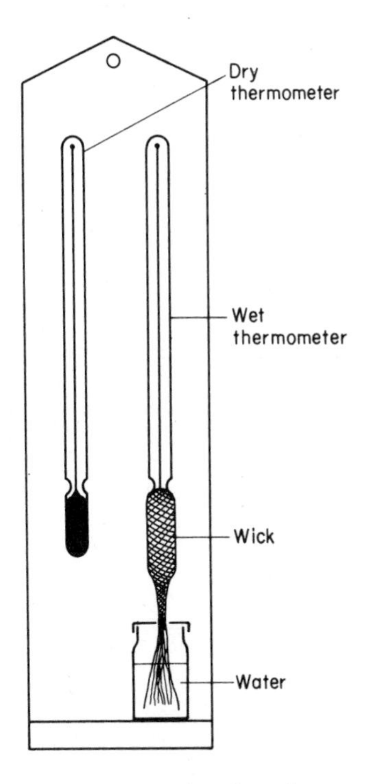

Fig. 31. A non-aspirated psychrometer.

psychrometers can give too high a value of relative humidity, and shoul therefore be used with caution. It is also particularly important that nor aspirated psychrometers are not exposed to direct solar radiation.

Aspirated psychrometers show great variation in design and the means c ventilation include electric or clockwork fans, whirling the thermometer around a handle and sucking through a plastic tube. In the Assmann psychro meter, which is a large and very accurate meteorological instrument, ventila tion is provided by a clockwork or electric fan mounted at the top of a pair o vertical tubes containing mercury thermometers. The bulbs of the thermo meters, one of which is fitted with a wick, are enclosed in a radiation shield a the lower end. The whirling psychrometer (Fig. 32) is a much simpler bu quite effective instrument, widely used by ecologists. In order to achieve a aspiration rate of 3 m/s, it has to be rotated at about 2 revolutions per secon (rev/s). Both the Assmann and whirling psychrometers use or disturb a larg volume of air in order to obtain a measurement and their use in the micro climate field is therefore limited. Miniature psychrometers that use electrica temperature sensors are more appropriate to small-scale measurement Several designs are available commercially and a simple design for an instru ment that uses less than 0.3 litres of air to obtain a reading will be describe later in this chapter.

B. Dew-point Method

If a sample of air that contains water vapour is cooled, its degree of satura tion will increase until it becomes fully saturated and any further cooling wil

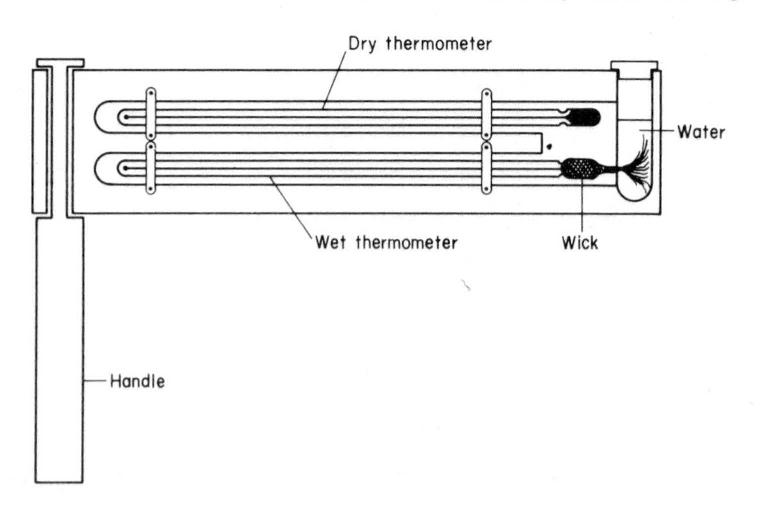

Fig. 32. A whirling psychrometer.

ause condensation. The temperature at which this occurs is called the dew-point and depends on the amount of water in the air. Dew-point is normally measured by cooling a surface to below the point of saturation, allowing water to condense onto it, and then gradually raising the temperature until he film of condensation starts to evaporate. The temperature at which this change occurs is taken as the dew-point temperature. The presence of the film of condensed water can be measured optically (Easty and Young, 1976), measured electrically or simply observed. Methods of cooling the surface ange from the use of electrical Peltier-effect junctions to the evaporation of a few drops of a volatile fluid such as ether. One of the most convenient of he non-automatic methods is the use of an aerosol can of refrigerant gas. Dew-point instruments are rather slow in operation but they can be used for reasonably small-scale measurements and give accurate estimates of humidity.

C. Change of Physical Dimensions

Many materials change their physical dimensions when they absorb water and this property can be used to make instruments that measure humidity. The simplest and least satisfactory instrument is the paper hygrometer which uses a spiral of paper, varnished on one side, as the sensitive element. The varnish stabilises the dimensions of the paper on one side while the other side changes with humidity, resulting in changes in curvature of the spiral. Paper hygrometers are badly influenced by temperature but the hair hygrometer is a much better instrument. Human hair becomes longer as the air becomes wetter and gets shorter as the air becomes drier and this property can be used in a simple hygrometer. Since the changes in length are small, some magnification of the change in length has to be provided. Hair hygrometers are also affected by changes in temperature, although to a lesser degree than paper hygrometers and an instrument adjusted to read correctly at 15°C will read 10% low at 38°C and 10% high at −6°C. Provided an allowance is made for the effect of temperature, hair hygrometers are usually accurate to within 5% over most of the range of humidity, but they should not be subjected to a relative humidity of less than 5% or a temperature exceeding 65°C and the calibration should be checked regularly.

D. Change in Concentration of a Solution

If a small droplet of an aqueous solution is exposed to the air, water will move in the appropriate direction between the solution and the air until equilibrium is reached. The concentration of the solution will be directly related to the relative humidity of the air. This technique can be used for very small-scale measurements, and will be described in greater detail later in this chapter.

E. Cobalt Thiocyanate Paper

Paper soaked in a cobalt thiocyanate solution and allowed to dry has the property that its colour changes from red to blue as the humidity decreases. The changes in colour are most marked in the range 50 – 90% relative humidity. Paper that has been exposed to the air should be quickly mounted in liquid paraffin over a white surface and covered by a piece of glass, and it can then be compared with a colour chart. Although the commercially prepared paper seems reasonably consistent, this is a rather approximate way of estimating humidity, but it can sometimes be useful in making comparisons between different measurement points. If one is asking questions such as "Is point A wetter, drier or about the same humidity as point B?" then cobalt thiocyanate paper is worth considering. It can be used in very small spaces.

F. Changes in Electrical Properties of Materials

When materials absorb water, their electrical properties are usually changed. The best example of a sensor that changes its resistance in response to changes in humidity is the sulphonated polystyrene layer type. These are rather slow in response, usually taking two or three minutes to reach their final reading, and they suffer from a certain amount of hysteresis (often ±3%). Their resistance is usually in the megohm region.

The lithium chloride hygrometer consists of a skein of very fine fibres impregnated with lithium chloride, wound around a pair of wires which are powered by an a.c. supply. Lithium chloride is a hygroscopic chemical that takes up water if the ambient relative humidity is greater than 11%. The resistance between the wires decreases as water is absorbed, which causes current to flow, raising the temperature and reducing the rate of absorption of water. When the system reaches equilibrium, the temperature of the sensor can be related to the dew-point temperature of the ambient air. Lithium chloride hygrometers are slow in response and can be damaged by prolonged exposure to saturated air.

Capacitance hygrometers measure the change in capacitance caused by the absorption of water into a dielectric, usually aluminium oxide. One method of construction involves anodising an aluminium substrate to produce a porous layer of aluminium oxide a few microns thick, which is covered by a very fine porous coating of gold. The substrate and the gold layer form the plates of a capacitor, with a porous aluminium oxide dielectric. The capacitance is measured with a high-frequency oscillator and bridge circuit. Capacitance hygrometers have a very quick response, usually only seconds, and can be reasonably accurate ($\pm 3\%$).

G. *Infra-red Gas Analyser*

The absorption of radiation in the long wavelength part of the infra-red band can be used to estimate atmospheric humidity. This technique will only be mentioned briefly since at present the equipment is very complex and unsuited for field use, although this may not always be the case. Infra-red gas analysers respond quickly, and can be very accurate. They are also used to measure carbon dioxide concentrations.

H. *Change in Resonant Frequency of a Quartz Crystal*

If a quartz crystal is coated with a very thin hygroscopic layer, the change in mass of this layer on absorption of water will cause a change in the resonant frequency of the crystal. If the crystal is connected to an oscillator circuit, the output frequency will be related to the humidity of the air. Sensors of this type are used in the chemical industry.

II. Major Sources of Error in Humidity Measurement

Humidity sensors usually modify the humidity nearby to some degree and they may also cause air movements and modify the temperature. The psychrometer takes a volume of air and evaporates water into it, thus increasing the humidity and reducing the temperature. The modified air is then discharged, either close to the measurement point or at a distance from it. If the modified air is discharged close to the measurement point, it may cause serious modification of the humidity and temperature unless there are adequate natural air movements to cause dispersal. Small aspirated psychrometers, which can be made to sample as little as one-third of a litre of air and which discharge the modified air well away from the measurement point, will only cause serious errors if the air which replaces that used by the instrument has a different temperature or humidity. Except where measurements are attempted in small enclosed areas, this will generally not be the case.

Dew-point sensors take water from the air in order to form a film of water on the dew-plate, thus reducing the humidity of the air in close proximity to the plate. When the temperature of the plate is raised, water will start to evaporate when the dew-point temperature of this air is reached. In the presence of even very small air movements, this effect will be of little importance, but in small enclosed spaces it can be a serious source of error. Another problem with dew-point determinations is the need to take measurements of air temperature, which are clearly likely to be affected by the close proximity of a cooled dew-plate. The use of a Peltier-effect thermoelectric cooler, which also has a hot junction, may further complicate matters. Similar considerations also apply to other types of humidity sensor that do not operate at ambient temperature (e.g. the lithium chloride device); none of these devices is capable of giving an instantaneous reading of relative humidity at a point. The two measurements involved, such as dew-point and temperature, must be taken in different places or at different times. However, provided account is taken of these limitations, dew-point determinations can be of great value, and allow humidity measurements to be made on quite a small scale.

The group of humidity sensors in which water is absorbed into a material, with consequent change in its properties, modify their environment to a much smaller extent. Hair hygrometers, capacitance sensors and the aqueous solution droplet technique can be used in almost all circumstances without significant errors due to water moving in or out of the material. In very small-scale measurements in enclosed areas there may be problems and if it is suspected that the sensor is seriously modifying its environment, successive readings can be taken with the sensor previously equilibrated to humidities above and below that being measured. Some devices in this group take a long time to reach equilibrium with the air. The equilibration time depends on the humidity, temperature and air movement. In still air, at high humidity and low temperature, some hair hygrometers may take as long as an hour to attain their final reading.

III. Simple Equipment for Humidity Measurement

A. Non-aspirated and Whirling Psychrometers

The non-aspirated psychrometer is illustrated in Fig. 31 and the whirling psychrometer in Fig. 32. These are very simple instruments which can be home constructed without difficulty although commercial instruments are available at moderate cost. In the non-aspirated psychrometer it is important that the thermometers are mounted well away from the back-plate to allow for good air circulation, and they should always be shielded from direct solar radiation. In the whirling psychrometer, it is necessary to anchor the thermometers securely to prevent them sliding towards the water container. Special thermometers can be obtained with the end remote from the bulb expanded into a flat disk to facilitate mounting in a whirling psychrometer. Special wicks can also be obtained for non-aspirated and whirling psychrometers although woven cotton boot-laces are also effective. To prevent contamination of the wick, only distilled water should be used in either instrument.

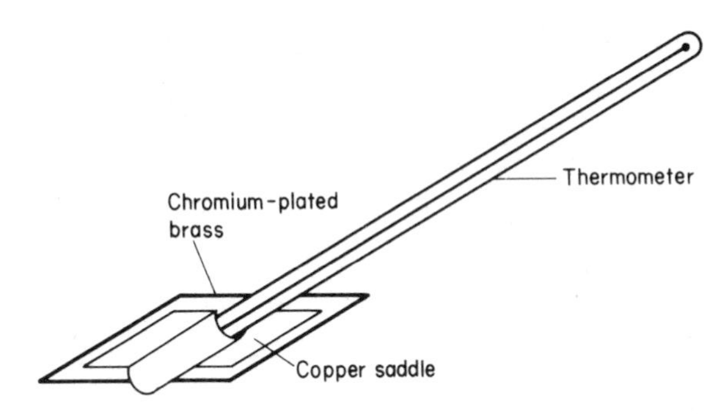

Fig. 33. Dew-point device with mercury thermometer.

B. Dew-point Method

Figure 33 shows a very simple dew-plate with its temperature indicated by a mercury thermometer. It is made from a small piece of chromium-plated brass (cut from a photographic glazing plate) with a copper saddle soldered to the non-plated side. A thermometer is fitted into the cavity so produced and either permanently secured with epoxy resin adhesive or a thermal grease used to improve thermal contact between the thermometer and the dew-plate. The compounds used for mounting transistors onto heat-sinks in the electronics industry are suitable. The dew-plate can be cooled by evaporating a few drops of ether onto the back surface until a film of dew forms, but a more convenient method is the use of an aerosol can of refrigerant gas (freezer aerosol, as used in the electronics industry for tracing cracks in printed-circuit boards). It should be noted that only a very small quantity of gas is needed and a too generous application will cause the formation of frost, which should not be confused with dew. Once the dew has formed on the plate, the instrument should be carefully placed in position and observed from a suitable distance so that the presence of the operator does not unduly affect the reading. It is usual to take the dew-point as the temperature at which half of the area of the dew-plate is clear.

A dew-point device with electrical indication is shown in Fig. 34. The temperature sensor can be a thermocouple, thermistor or silicon diode and may be used with the appropriate indicator described in Chapter 1. This instrument is intended to be cooled with refrigerant gas (freezer aerosol) which is squirted into one of the nylon tubes, exhausting through the other. The gas passes through holes drilled in a copper block (20 × 20 × 4 mm) to which a piece of chromium-plated brass has been soft soldered. The temperature sensor is inserted into a smaller hole drilled into the centre of the block, as shown in the section in Fig. 34a. Because of the greater thermal mass of

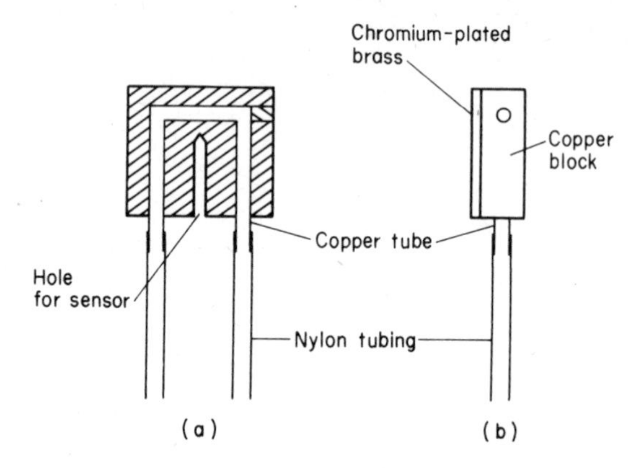

Fig. 34. Dew-point device with electrical sensor.

this device, the temperature will rise more slowly than in the previous design, with consequent improvement in accuracy.

Care should always be taken to keep the surface of any dew-plate clean and free from grease, although it is sometimes found that a small application of saliva will help in the formation of an even film of dew.

C. *Miniature Electric Psychrometer*

This instrument uses small thermocouples as the temperature sensors and is aspirated by simply sucking through a polythene tube. Because the dimensions of the measurement tube containing the thermocouples are small, an

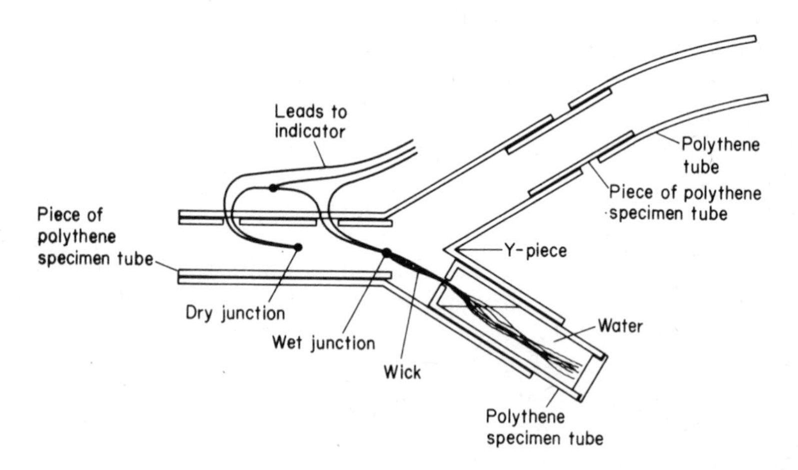

Fig. 35. Section through the head of the miniature electric psychrometer.

adequate rate of aspiration can be maintained for the few seconds needed for a determination without undue effort. The psychrometer shown in section in Fig. 35 was developed from an earlier version (Unwin, 1979) but uses easily available laboratory polythene-ware in place of the machined Perspex components. The psychrometer head is built in a polythene Y-piece which has an internal diameter of about 1 cm. A polythene specimen tube with a snap-on lid serves as the water container, and is a push-fit into one of the arms of the Y-piece. A small hole is made in the bottom of the tube for the wick of the wet junction. A second specimen tube, with its bottom and top removed, is inserted into the main stem of the Y-piece to reduce the diameter, and a third specimen tube with its ends removed is used to join the remaining arm of the Y-piece to the aspiration tube (right-hand side of Fig. 35).

The wick of the wet junction is made by passing three lengths of cotton sewing-machine thread over the thermocouple junction (Fig. 36 a) and binding them, not too tightly, with another length of the same thread. The six ends of the threads which pass over the junction and the two lengths of the binding thread form the wick (Fig. 36 b). The junctions are made from 36 swg copper and constantan wire. The constantan leads from the wet and dry junctions are joined together and are connected, with the two independent copper leads, to the indicator. The fine wire from which the thermocouples are made would be too fragile to form a connecting cable and should therefore be joined to thicker cables of the same material, observing the precautions outlined in Chapter 1.

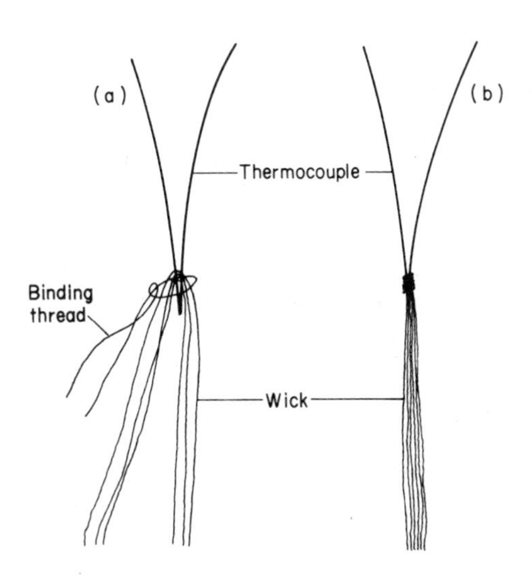

Fig. 36. Construction of the wet junction.

The miniature electric psychrometer may be connected to any thermocouple indicator, such as that shown in Fig. 10. The circuit given in Fig. 37, which was specially designed for use with this psychrometer, combines the thermocouple indicator of Fig. 10 with the thermistor meter of Fig. 3, used as a reference thermometer. The thermistor (T) is mounted in an insulated aluminium block, adjacent to the reference thermocouple (RJ). The master switch (S1), which is a four-pole, three-position wafer switch, is used to control the functions of the instrument. In position 1 the instrument is switched off, in position 2 the meter reads the reference temperature and in position 3 it functions as a thermocouple meter and the input switch (S2) becomes operative. With the input switch (S2) in position 1, the input of the amplifier is short-circuited and the meter reading should be set to equal the reference temperature (read previously with S1 in position 2) by adjusting the 10-turn helical potentiometer (Cal). With S2 in position 2, the dry bulb temperature will be indicated, and in position 3 the wet bulb temperature. Position 4 connects the meter to an external thermocouple probe. The pre-set adjustments are as follows: T sets the range of temperatures covered by the meter scale in the thermocouple mode, S sets the range covered in the reference mode and Q sets the zero point in the reference mode. The thermistor

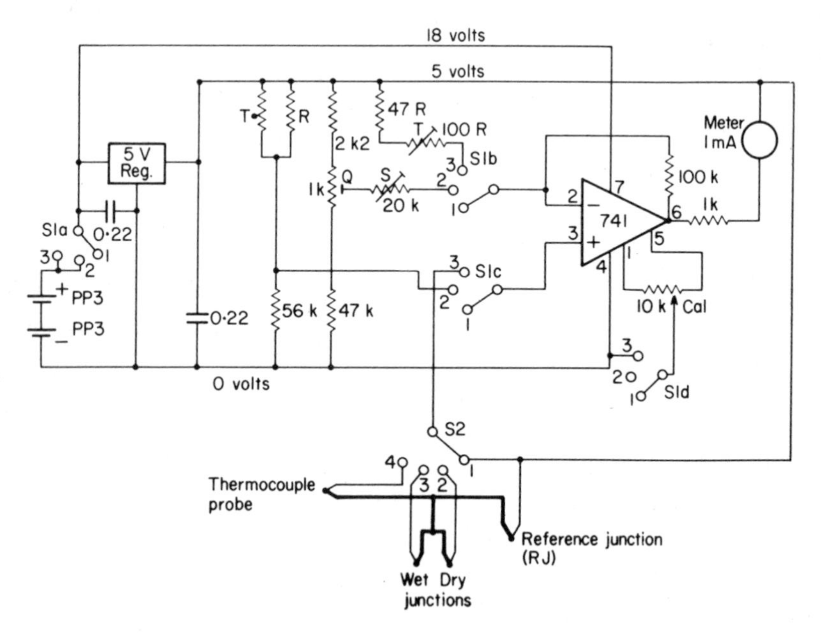

Fig. 37. Circuit diagram of the indicator for the miniature electric psychrometer. Thickened lines indicate constantan leads.

shunt resistance (R) should be selected to linearise the thermistor characteristics, as described for the thermistor meter in Chapter 1.

As with all psychrometers, only distilled or de-ionised water should be used, to avoid deposits on the wick.

D. Corbet Aqueous Solution Droplet Technique

As was stated earlier, the concentration of an aqueous solution in equilibrium with the air is related to the relative humidity. In this technique, very small droplets of a suitable solution are exposed to the air and allowed to equilibrate. Their concentration is then measured in a pocket refractometer that has been specially modified by the makers (Bellingham and Stanley, Tunbridge Wells, Kent, England) to operate on small samples ($< 0.5\ \mu\ell$). The volume of the drops is about 1 $\mu\ell$ when loaded to allow for a reduction in volume at low humidities. The equilibration time will depend on factors such as the humidity, temperature and air movement but is often in the order of 30 minutes. Probably the best technique is to use a number of replicate drops and to measure their concentration at, say, 5 minute intervals to be certain that equilibration has taken place. This technique was originally developed for measuring the humidity of the air inside flowers (Corbet *et al.*, 1979a) and the droplets were held in loops in a "wiggly wire", made by winding 36 swg epoxy-coated copper wire around size 00 (0.4 mm diameter) entomological pins, inserted into a piece of wood at 5 mm intervals and cut off at a height of about 5 mm. This produces the form shown in Fig. 38a. Another method is to use a strip of very fine nylon net with a few of its holes filled with the solution (P.G. Willmer, personal communication), as shown in Fig. 38b. In either case, droplets are unloaded and transferred to the refractometer using microcapillary tubes. The choice of solute will depend on the range of humidities being measured, but for the range 40 – 100%, potassium acetate is suitable. For high humidities, particularly over 90%, sucrose will give better definition. The characteristics of both solutes are given in Fig. 39.

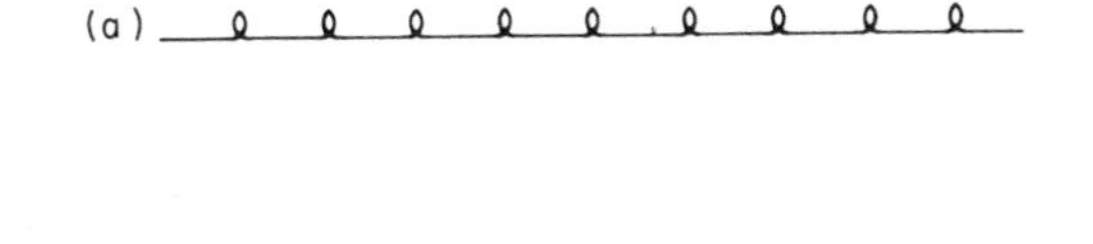

Fig. 38. Two methods of holding drops of solution: (a) "wiggly wires" and (b) strips of nylon mesh.

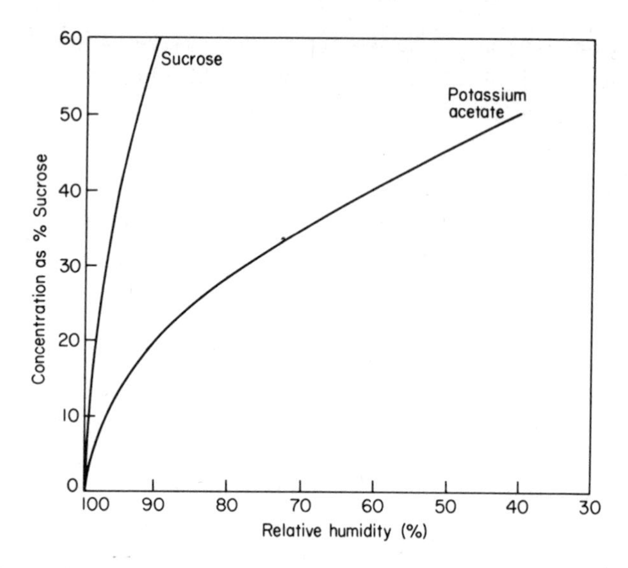

Fig. 39. Humidity/concentration characteristics for potassium acetate and sucrose.

This is a very powerful technique, enabling measurements to be made on an extremely small scale, with the minimum of equipment. The amount of water which moves into or out of the droplet is very small and the technique can be used in very small enclosed areas.

E. Electrical Conductivity of Aqueous Solution

In the aqueous solution droplet technique, the refractive index was used to estimate the concentration of the solution, but it can be measured electrically. If a very small droplet of a suitable solution, such as potassium acetate, is held between a pair of platinum ring electrodes, the electrical impedance can

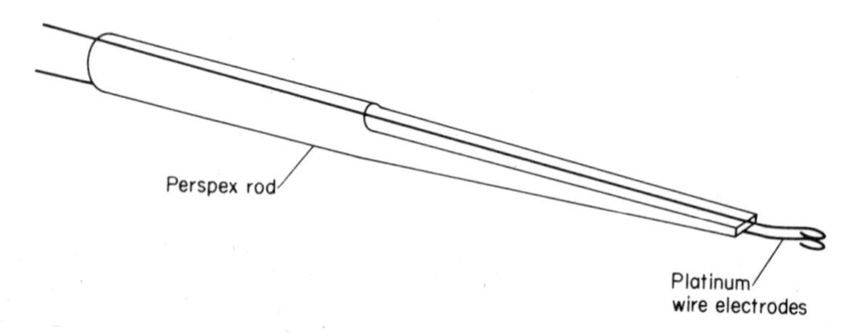

Fig. 40. Construction of the electrode.

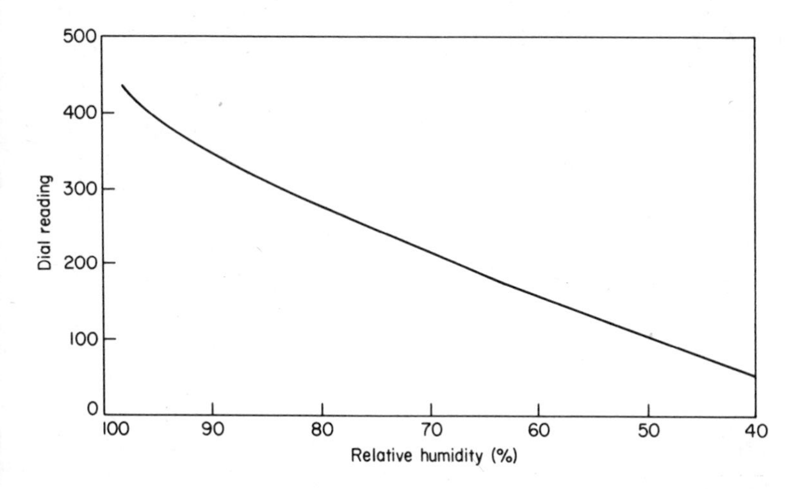

Fig. 41. Characteristics of a potassium acetate droplet in the prototype electrode.

be measured in a bridge circuit. The design of the electrode is shown in Fig. 40. The ring electrodes are made from 0.2 mm (0.005 inch) platinum wire which is formed into a loop around a piece of 22 swg tinned copper wire and pulled tight. One end is then cut from the loop and the wire formed into the shape illustrated. The two electrodes are then cemented to flats filed on opposite sides of a piece of 3 mm Perspex rod with Perspex cement. The separation between the loops is approximately equal to the wire diameter.

The droplet of solution held between the electrodes has a very small volume (about 0.3 $\mu\ell$) and equilibrates rather more quickly than in the previous technique (usually a stable reading is obtained within 7 minutes). The volume held by the electrodes is determined by their geometry and provided the same

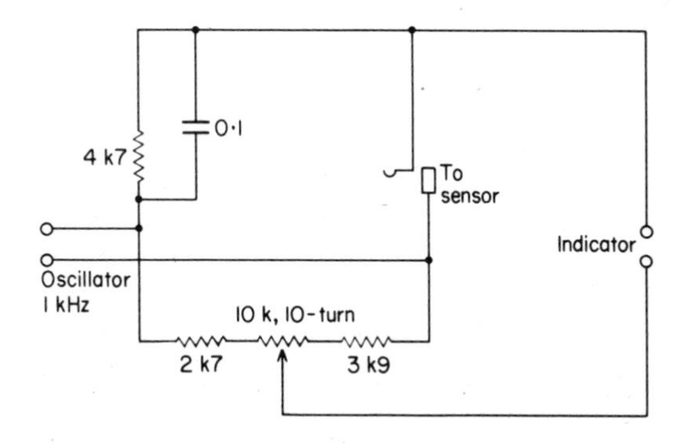

Fig. 42. Simple bridge circuit for measuring conductivity of a droplet of potassium acetate solution.

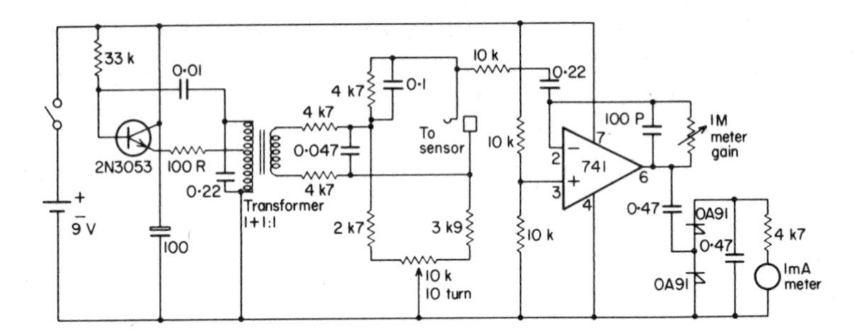

Fig. 43. Circuit diagram of a complete bridge and indicator for measuring the conductivity of droplet of potassium acetate solution.

concentration of solution is always used, remarkably consistent results ar obtained. The prototype electrode was charged with a concentration of 20% and the characteristics are given in Fig. 41.

A simple bridge circuit for use with this technique is given in Fig. 42. Th oscillator is an audio frequency generator with an output of about 1 volt at kilohertz (kHz). It should have a "floating" output or be isolated fron ground (any battery-driven oscillator will do). The indicator may be an oscil loscope or an a.c. millivoltmeter. The 10-turn helical potentiometer, whicl should be fitted with a counting dial, is adjusted for minimum output on th indicator. A more complex circuit, incorporating an oscillator, bridge anc indicator, is given in Fig. 43.

This is one of the simplest ways of obtaining a remote electrical indicatio of humidity. It does not suffer from the problem of sensor contaminatio over a long period, which occurs in most resistive humidity sensors, because i can be regularly cleaned and recharged with solution. It permits very small scale measurements to be made, even in sealed containers, and cause minimal modification of its environment.

IV. Calibration and Testing of Humidity Sensors

Humidity sensors can be calibrated or tested in two ways, by the use of "humidity standards" or by comparison with instruments of known performance. Humidity standards may be set up by the use of solutions of sulphuric acid or potassium hydroxide, or by the use of saturated salt solutions (Solomon, 1951; Winston and Bates, 1960; Cloudsley-Thompson, 1967; see also Table 3 in Chapter 5). These techniques for the control of humidity are covered in detail in Chapter 5. In all cases, care should be taken that the vessel containing the air sample and the solution is well sealed, that the temperature

not allowed to vary, and that enough time has been allowed for equilibra-
on to take place. In some cases, the equilibration time can be very long
ndeed, although the use of a pump to take air from the top of the vessel
nd bubble it through the solution may considerably shorten the process.
However, it is always better to have an independent means of estimating the
umidity of these "standards"; even if the independent method is of only
moderate accuracy, it will indicate when equilibration has taken place and
void the more serious errors. Perhaps the most serious limitation of the use
of humidity standards of this type is that they cannot be used to test or
alibrate any of the humidity sensors that change their local environment by
absorbing or evaporating water (e.g. dew-point or psychrometer) or introducing temperature gradients (lithium chloride or dew-point), without
introducing the possibility of serious error. These methods require a reasonable amount of air movement for accurate operation.

The safest method of calibration of any type of sensor is to compare it with
a good instrument of known performance under realistic conditions. A range
of humidities of 50 – 95% RH can often be found between mid-day and
mid-night on a summer day, and other methods, such as running the shower
n a domestic bathroom, can be used for the high range. Humidities below
0% can sometimes be found in centrally heated offices and can be artifi-
cially created by the use of electric heaters indoors. The relative humidity of a
room which is normally 60% at 20°C may be reduced to 30% by raising the
temperature to 32°C. The best instruments to use as standards are the
psychrometer and dew-point apparatus. A dew-point apparatus with
accurately calibrated thermometer, and an equally accurate measurement of
air temperature, will give good estimates of humidity. At a much higher cost,
the Assmann psychrometer is ideal but the whirling psychrometer is probably
accurate enough for most purposes. There is a great deal to be said in favour
of the use of the simple whirling psychrometer as a basic standard humidity
instrument against which other, more specialised instruments are tested and
calibrated.

4
Wind

I. The Measurement of Wind Speed and Direction

Of all the elements of microclimate, wind is the most transitory. On some occasions the velocity can be almost constant, while on others it may vary by an order of magnitude over a short period. Only a continuous recording of the instantaneous wind speed measured in three directions, mutually at right angles, can give a complete picture of air movements. Fortunately, for many purposes, such detailed measurements are not necessary and it is often sufficient to express wind speed as a mean value over the measurement period together with the maximum value recorded during the same period. However, it should always be remembered that air movements are three-dimensional and transitory.

A. Wind-vanes

Simple wind-vanes of the "weathercock" type indicate wind direction in the horizontal plane, and differ mainly in the shape of the fin. Some designs, usually called trivanes, also align themselves with the vertical angle of the wind. Wind-vanes may directly indicate the wind angle on scales fixed to the instrument, or they may incorporate a transducer and indicate the angle remotely.

B. Pressure Anemometers

These are usually of the Pitot-tube type (Fig. 44). The pressure in an open tube facing into the wind is compared with that of a "static vent", usually with a manometer. The pressure difference is related to the wind speed as follows:

$$\text{Velocity (m/s)} = 126\sqrt{h}$$

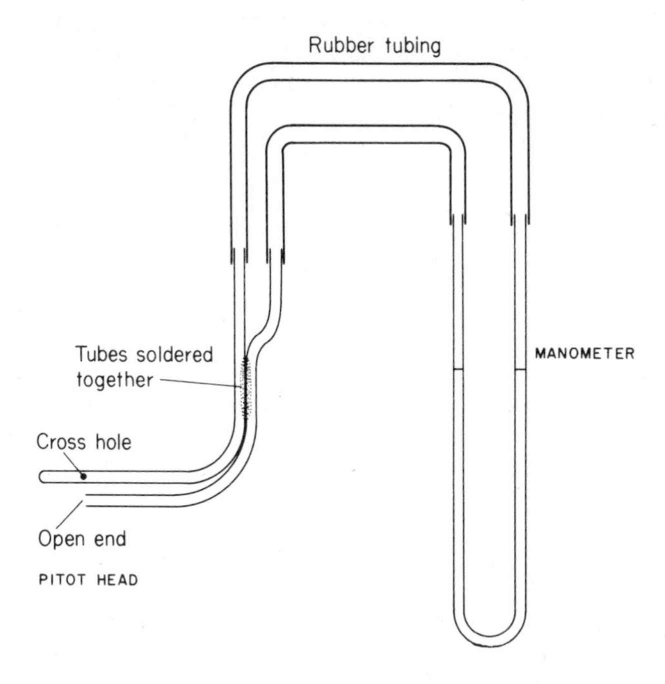

Fig. 44. Pitot-tube anemometer with manometer.

where h is the pressure difference in metres of water. This pressure difference is too small for estimating the low speeds encountered in microclimate work (1 cm of water is equivalent to 12.6 m/s), but the Pitot-tube anemometer can be valuable at higher speeds, particularly for calibration purposes.

C. Mechanical Anemometers

These are commonly of the cup anemometer or the propeller type. The cup anemometer usually consists of three arms carrying approximately hemispherical cups which present a greater drag when the concave side faces the airstream. The speed of rotation of the cup assembly may be indicated directly, usually as metres per second or kilometres per hour, and may have provision for recording the maximum speed achieved. Alternatively, the number of rotations of the cup assembly, suitably modified by gearing, may be indicated as metres or kilometres, and this value should be divided by the observation time to calculate the mean wind speed. Cup anemometers are not affected by changes in wind direction in the horizontal plane and do not respond to vertical components of air movement.

Propeller-type anemometers are more directional than cup anemometer and are either aligned to the wind direction, usually by mounting them on a wind-vane, or mounted in a set of three, arranged mutually at right-angles, to indicate the components of the wind in three directions (usually N – S, E – W and vertical).

D. *Thermoelectric Anemometers (Thermoanemometers)*

Wind speed can be estimated by measuring the rate of conductive cooling of a heated body. The sensing element is usually a hot wire or a thermistor, and the output is either the current required to achieve a predetermined tempera- ture above ambient, or the temperature difference from ambient resulting from a fixed current flow. Thermoanemometers can be made to be very insensitive to wind direction, or they may be fitted into tubes that are mounted mutually at right-angles for directional measurements.

E. *Sonic Anemometers*

These are a recent development and rely on the fact that the apparent speed of sound in air, measured between fixed points, depends on the rate at which the air is moving. Three sets of transmitters and receivers, operating on different frequencies, are set up mutually at right angles, and the three-dimensional components of air movement are measured simultaneously. These instru- ments are inevitably very complex, but have by far the quickest response of any type of anemometer.

II. Major Sources of Error in Wind Measurement

Mechanical anemometers, such as the cup or propeller type, have a threshold below which the friction of the system prevents rotation. Their inertia makes them slow to start, and causes them to overrun when the wind speed drops. Mechanical anemometers should therefore have inertia and friction as low as possible. Weather station instruments have to be robust in construction, which generally results in rather high inertia and friction, but miniature devices intended for short-term use, although less robust than the larger instruments, can be made of light materials, and can have very free bearings. Such instruments can be made to respond to wind speeds down to 0.1 m/s. The miniature cup anemometer is a very useful instrument for measuring average wind speeds.

Thermoanemometers usually have an output which is proportional to the logarithm of the wind speed, which means that they cannot be fed directly into an integrator to estimate the mean speed. The speed of response usually varies with the wind speed, having a faster response at high speeds. This means that short gusts have a better chance of being faithfully recorded if the velocities involved are high.

Perhaps the greatest uncertainty in wind measurement is the problem of how to express, in a simple way, a quantity with a very large variance.

III. Simple Equipment for Wind Measurement

A. Wind-vane

A simple direct-indicating wind-vane is shown in Fig. 45. The 70 × 80 mm × 18 swg aluminium fin is fixed into a slot cut into a 5 mm aluminium alloy rod with epoxy resin adhesive. The rod is glued into a 5 mm hole drilled through the central pivot, at a distance of 130 mm from the fin. In the prototype, the balance weight, which was made from 13 mm diameter brass, 40 mm long, was drilled with a 5 mm hole and fixed to the aluminium rod at a distance of 45 mm from the pivot. The exact distance will depend on the balance of the vane and should be adjusted for static balance with the axis of the pivot horizontal. One half of a table tennis ball is used to provide a rain shield for the ballrace and to form a calibrated scale. The method of mounting of the ballrace will depend on the facilities available to the constructor, and if a lathe is not available, the simplest method is to choose a bearing that exactly fits the pivot shaft, which is 10 mm in diameter. It can be secured with epoxy adhesive. The handle can then be drilled out to fit the outer part of the ballrace.

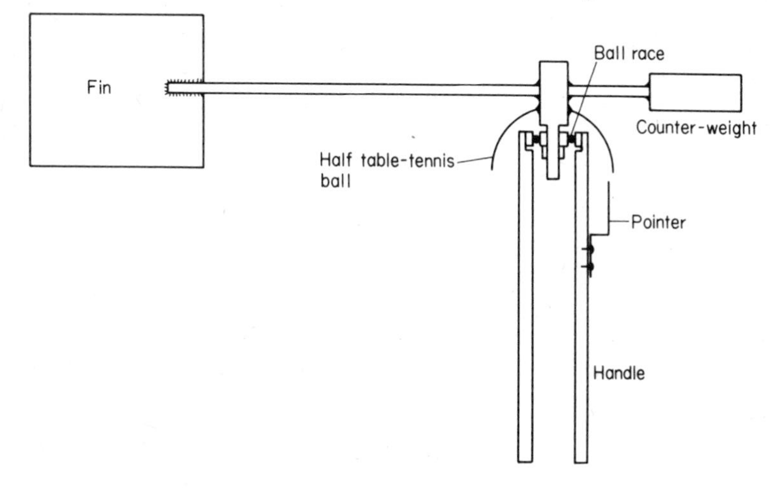

Fig. 45. Simple wind-vane.

B. Pitot-tube Anemometer

A Pitot-tube anemometer is shown in Fig. 44, connected to a manometer. The Pitot-head is made from 5 mm diameter copper tubing normally used for microbore central heating systems. The lower, shorter tube has an open end and faces into the wind. The upper tube has a closed end and extends beyond the open end of the lower tube. It has two cross-holes, 1 mm in diameter, at a point level with the end of the lower tube, which form the static vent. The upper ends of the two tubes are connected to a water manometer. As was stated earlier, the pressure difference is related to the wind speed by the formula:

$$\text{Velocity (m/s)} = 126\sqrt{h}\ (h \text{ in metres of water})$$

C. Cup Anemometer

The mechanical arrangement of the cup anemometer, simplified by showing only one of the arms, is illustrated in Fig. 46. The rotation of the cup assembly causes the dry reed switch to open and close when the field from the magnet reverses in direction, thus generating pulses which are amplified and counted on an electromechanical counter. By using a powerful magnet, the reed switch can be mounted well away from it and the differences in "pull" on the magnet with changes in position are minimised. The cups are made from halves of table tennis balls, glued into loops made in lengths of 20 swg tinned copper wire. The wire should first be straightened by stretching it to just beyond its elastic limit, and a loop made in the centre by winding it around a cylindrical object about 20% smaller in diameter than the table tennis balls. When released, it will spring out to the diameter of the ball and the ends can be shaped as shown in Fig. 46 with long-nosed pliers. The arms

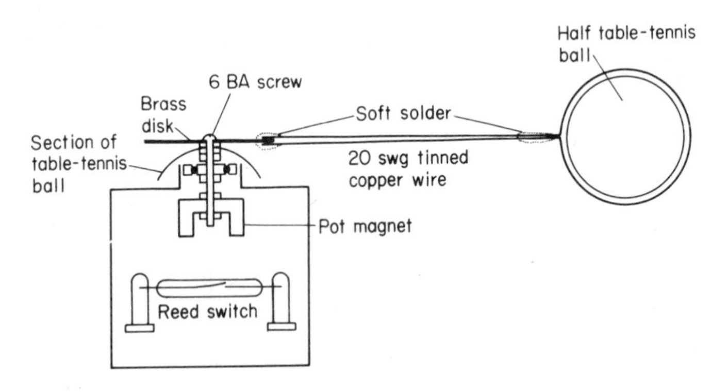

Fig. 46. Mechanical arrangement of cup anemometer.

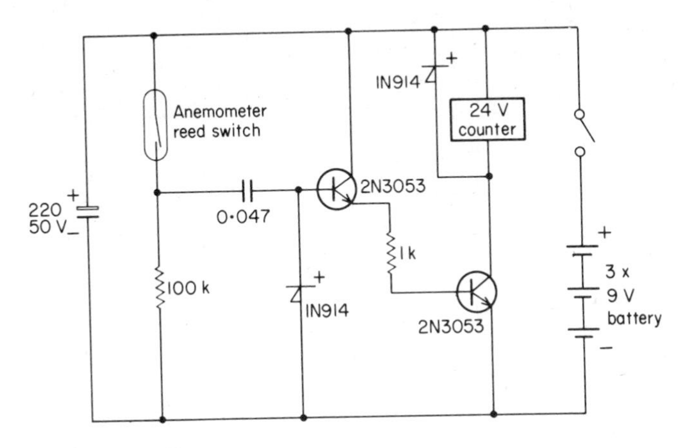

Fig. 47. Circuit diagram of cup anemometer indicator.

should be about 90 mm in length, measured from the hole in the centre of the brass disk to the inside edge of the cup. Longer arms will improve performance at low speeds, but will increase inertia while shorter arms will be more likely to suffer from the effects of friction at low speeds. The ballrace should be small in diameter, and in the prototype had an outside diameter of 12 mm and a 3 mm hole. The ballrace is protected from rain by a section of a table tennis ball, fitted onto the pivot shaft just below the brass disk. The pivot shaft is a screw 3 mm in diameter or equivalent (6 BA).

The circuit of the indicator is given in Fig. 47. The closing of the reed switch causes a pulse of current to be fed through the 0.047 microfarad (μF) input capacitor into the transistor amplifier. The pulse has been made just long enough to operate the counter reliably (in order to minimise battery current) but with some counters the value of this capacitor may have to be increased.

This is a very free-running anemometer with low inertia. If elaborate workshop facilities are available to the constructor, the rotor assembly could be made from machined aluminium components, but care should be taken that the weight is kept to a minimum.

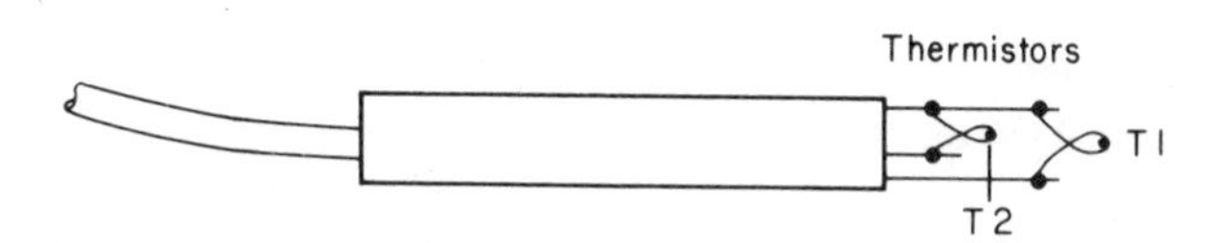

Fig. 48. Thermoanemometer probe.

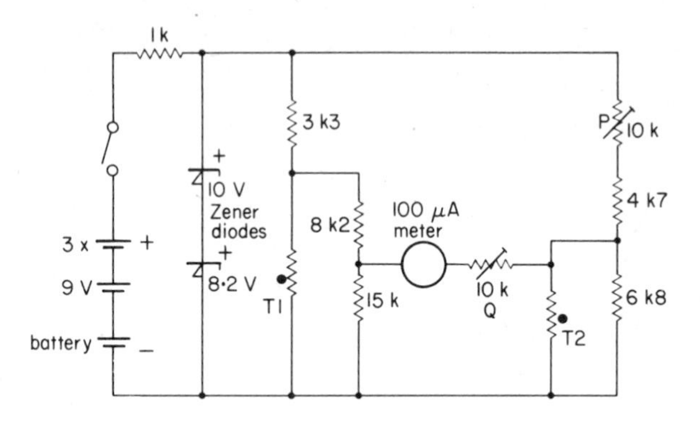

Fig. 49. Circuit diagram of the thermoanemometer. T1 and T2 are 4k7 thermistors type GM472 or VA3404 (R.S.Components Ltd, 151 – 142).

D. Thermoanemometer

This design uses a pair of miniature thermistors, mounted at the end of a 10 mm diameter probe, as shown in Fig. 48. The circuit (Fig. 49) is arranged so that the power dissipated in thermistor T1 is high enough to cause the temperature to rise well above ambient in still air. This temperature excess is dependent on the amount of convective cooling, and consequently on the wind speed. If the temperature indicated by the "overheated" thermistor (T1) is compared with the ambient temperature, indicated by the thermistor (T2), the difference will be approximately proportional to the logarithm of the wind speed. In a practical circuit, the thermistor measuring ambient temperature will also run slightly above ambient (see Chapter 1) and its own temperature excess will be dependent on wind speed, but provided the power dissipated in the "overheated" thermistor (T1) is considerably greater than that dissipated in the "reference" thermistor (T2), the performance of the instrument will not be seriously affected. The pre-set potentiometer (P) sets the zero point and Q sets the range of speeds covered by the meter (0.05 – 20 m/s in the prototype). Care should be taken when setting the zero point that the air speed really is zero, since very small air currents will register on this type of anemometer. The probe should therefore be covered, ensuring that the cover does not touch the thermistors.

IV. Testing and Calibration of Anemometers

At low wind speeds, it is often easier to move the sensor at a constant speed than to attempt to move the air. Speeds of up to 3 m/s can be provided by fixing the sensor to the end of a pivoted arm and rotating at a suitable speed. Care should always be taken at very low speeds that the air in the room is not itself moving: a fixed thermoanemometer of the type described in this chapter will indicate if this is happening. At higher speeds, other methods will have to be used, but one should resist the temptation to use a motor vehicle. The air speed near to the body of a car is often considerably higher than the true speed, because of aerodynamic effects. A bicycle may be used in a long corridor, provided the sensor is held out in front on a pole, a procedure not without its hazards. If a wind tunnel is available, a Pitot-tube anemometer can be used as a standard.

5
Miscellaneous Topics

I. Control of Temperature

It is much easier to control temperature by putting heat into a system than by taking it out. Heaters respond much more quickly than refrigerators, and the life of most refrigerators is shortened if they are switched on and off frequently. If an artificial environment is required to operate at a temperature below ambient, with close control of temperature, it is usually better to cool it by a crudely controlled refrigerator to below the operating temperature, and use a heater to obtain fine control.

The simplest way of controlling electric power to a heater is by means of a thermostat. Thermostats usually consist of a pair of contacts which are mechanically operated by a temperature sensor. The sensor may be a bimetallic strip or a diaphragm moved by the pressure developed by a volatile fluid. Simple thermostats are often very slow in operation, due to the large sensors needed for direct operation of the contacts, and this can lead to inaccurate control, particularly in small enclosures. Electronic thermostats, which usually employ thermistors to operate a relay (electrically operated switch) through an amplifier, have very small sensors and can be quick and accurate.

The circuit of an electronic thermostat is given in Fig. 50. The voltage developed across the thermistor (T), which will depend on the temperature, is compared by the operational amplifier with that developed across a combination of a fixed (470R) and variable resistor (R), the total value of which sets the temperature. Since the temperature co-efficient of thermistors is negative, a fall in temperature will cause the voltage across the thermistor (T) to rise, and when it exceeds that developed across R and the 470R resistor, the output of the amplifier will go positive, turning on the transistor and operating the relay. The contacts of the relay (not shown) switch on the heater.

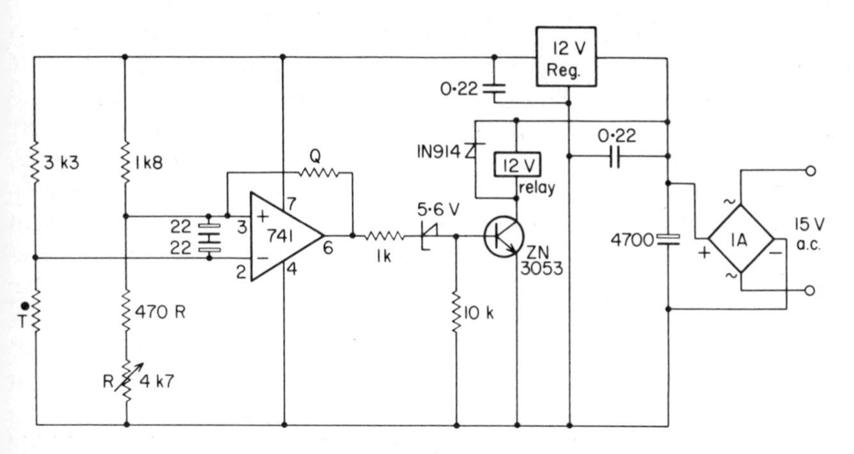

Fig. 50. Circuit of the thermistor temperature controller. T is a 4k7 thermistor. The circuit should be supplied from a suitable transformer (R.S.Components Ltd, 207 – 655).

When the temperature rises, causing a fall in the voltage developed across the thermistor, the process will be reversed, and the relay contacts will open, turning off the heater. The difference in temperature between the point at which the heater is turned on and that at which it is turned off (the differential) can be set by the value of the resistor Q. A value of 1 M gives a differential of about 0.3°C, and lower values give a larger differential. This circuit uses a 4k7 miniature thermistor and is capable of very accurate temperature control. Its temperature setting potentiometer (R) should be of the 10-turn helical type, which can be fitted with a lockable counting dial.

A very much simpler electronic thermostat is shown in Fig. 51. This circuit

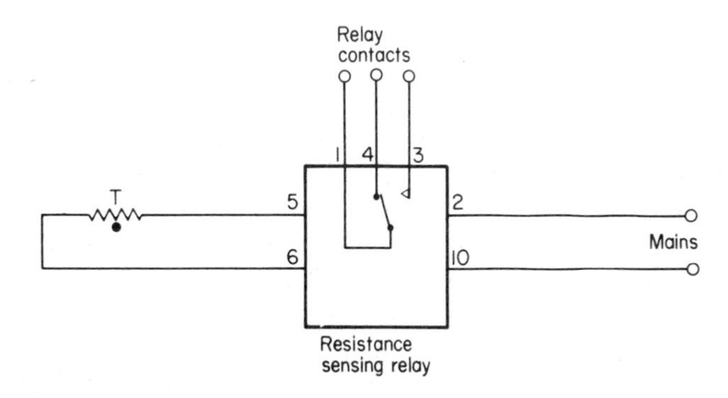

Fig. 51. Use of a resistance sensing relay for temperature control (R.S.Components Ltd, 349 – 822).

uses the same thermistor as the previous design, but connected to a resistance-sensing relay which directly controls the heater. Although the circuit would appear to be simple, the internal circuitry of the resistance-sensing relay is similar to that of Fig. 50 and the device is at present considerably more expensive than the total cost on components used in that design. However, the extreme simplicity of construction is very attractive, and recent experience with the electronics industry would suggest that the cost of this type of component is likely to fall rapidly. The relay is fitted with an internal potentiometer with which the thermistor's resistance is compared.

The type of heater used will depend on the application. The best heater for use in water is usually a standard immersion heater. For the control of air temperature in larger enclosures, a fan heater is often a useful device, but it should be rewired so that the fan runs continuously, and only the heater is controlled by the thermostat. It is unrealistic to assume that the temperature of any fluid is constant unless it is continuously stirred.

II. Control of Humidity

It is much easier to make air wetter than drier. On a large scale, one can use humidifiers which spray very fine droplets of water into the air. These can be controlled by a humidistat, which usually consists of contacts operated by a material that changes its dimensions with change in relative humidity, such as hair. Air-conditioning systems dry air by cooling it to well below the dew-point, when most of the water will condense out, and then warming it to the required temperature and re-humidifying it to a predetermined level.

On a small scale, water can be taken out of the air by use of hygroscopic materials such as silica gel. The humidity can also be controlled by allowing the air to equilibrate with an aqueous solution. Suitable solutions for this technique are potassium hydroxide and sulphuric acid, and some values for concentration and equivalent relative humidities are given in Table 3. A list is also given by Cloudsley-Thompson (1967). Since water will move between the air and the solution, the concentration of the latter will change and it is therefore good practice to ensure that the volume of the solution is relatively large, and to check the concentration from time to time.

A method of avoiding the problem of changing concentration is to select solutions that give the required relative humidity when saturated. The solution can be made up with an excess of solute so that a considerable amount of water can be taken up by the solution before any change in concentration takes place. Lists of saturated solutions and their equivalent relative humidities were given by Solomon (1951), Winston and Bates (1960) and Weast (1979). A selection of these is included in Table 3.

There are precautions that should be observed when using any of the

Table 3

Selection of Solutions for Control of Humidity

Solute	Concentration (g/100g sol.)	Specific gravity	Temperature (°C)	Relative humidity (%)	Deviation at 10 and 30°C (%)
KOH	64	1·55	20 – 25	10	
$LiC\ell.H_2O$	Saturated	–	20	12·5	+1, −1
KOH	50	1·51	20 – 25	15	
H_2SO_4	61	1·50	20 – 25	15	
KAc	Saturated	–	20	20	+1, −1
KOH	47	1·48	20 – 25	20	
H_2SO_4	58	1·47	20 – 25	20	
KOH	45	1·45	20 – 25	25	
H_2SO_4	55	1·45	20 – 25	25	
KOH	42	1·43	20 – 25	30	
H_2SO_4	52	1·42	20 – 25	30	
$MgC\ell_2$	Saturated	–	20	33	+1, −0·5
KOH	40	1·40	20 – 25	35	
H_2SO_4	50	1·40	20 – 25	35	
KOH	38	1·38	20 – 25	40	
H_2SO_4	48	1·37	20 – 25	40	
$K_2CO_3.2H_2O$	Saturated	–	20	44	+3, −0·5
KOH	36	1·35	20 – 25	45	
H_2SO_4	45	1·35	20 – 25	45	
KOH	34	1·33	20 – 25	50	
H_2SO_4	43	1·33	20 – 25	50	
$Mg(NO_3)_2.6H_2O$	Saturated	–	20	55	+3, −3
KOH	32	1·31	20 – 25	55	
H_2SO_4	41	1·31	20 – 25	55	
KOH	30	1·29	20 – 25	60	

Table 3 (*Continued*)

Solute	Concentration (g/100g sol.)	Specific gravity	Temperature (°C)	Relative humidity (%)	Deviation at 10 and 30°C (%)
H_2SO_4	38	1·29	20 – 25	60	
$NaNO_2$	Saturated	–	20	65·5	?, −2·5
KOH	27	1·26	20 – 25	65	
H_2SO_4	36	1·27	20 – 25	65	
KOH	25	1·24	20 – 25	70	
H_2SO_4	33	1·24	20 – 25	70	
KOH	22	1·21	20 – 25	75	
H_2SO_4	30	1·22	20 – 25	75	
NaCℓ	Saturated	–	20	76	+0·5, −0·5
KOH	19	1·18	20 – 25	80	
H_2SO_4	27	1·19	20 – 25	80	
KBr	Saturated	–	20	84	+1, −2
KOH	16	1·15	20 – 25	85	
H_2SO_4	23	1·16	20 – 25	85	
K_2CrO_4	Saturated	–	20	88	?, −1·5
KOH	12	1·11	20 – 25	90	
H_2SO_4	18	1·12	20 – 25	90	
$ZnSO_4.7H_2O$	Saturated	–	20	90	+3, −3
KNO_3	Saturated	–	20	93·5	+2·5, −2·5
KOH	7	1·06	20 – 25	95	
H_2SO_4	11	1·07	20 – 25	95	

KOH and H_2SO_4 can be considered to have a tolerance of ± 2% maximum over the range 0 to 40°C.

echniques involving aqueous solutions. Since the enclosure is presumably at relative humidity different from that of the air outside, water will move in or out unless it is well sealed. Opening the enclosure will change the humidity inside for a considerable period and should therefore be avoided if possible. n this connection, the microbalance described in the next section, which can be operated remotely without opening the chamber, can be particularly useful. Some of the saturated solutions have equilibrium humidities that are ignificantly affected by temperature and should only be relied upon to produce the correct humidity at a temperature at which the calibration is known. Although many of the solutions are not very sensitive to temperaure, the values for humidity given in Table 3 assume that the air and the solution are at the same temperature. Serious errors will result if there are temperature gradients in the system. For example, a solution that should control the relative humidity to 60% and which has a temperature of 20°C will produce a relative humidity of 68% if the temperature of the air is 2°C below that of the solution. It is also necessary that both the solution and the air are stirred, and a useful technique is to take air from the top of the chamber and bubble it through the solution using a small aquarium pump.

III. Microbalance

This is a design for a simple balance based on a moving-coil meter, which can be used in the field, and which can be remote-controlled for weighing inside a sealed chamber. The device used is a 1 mA panel meter of the "centre-pole"

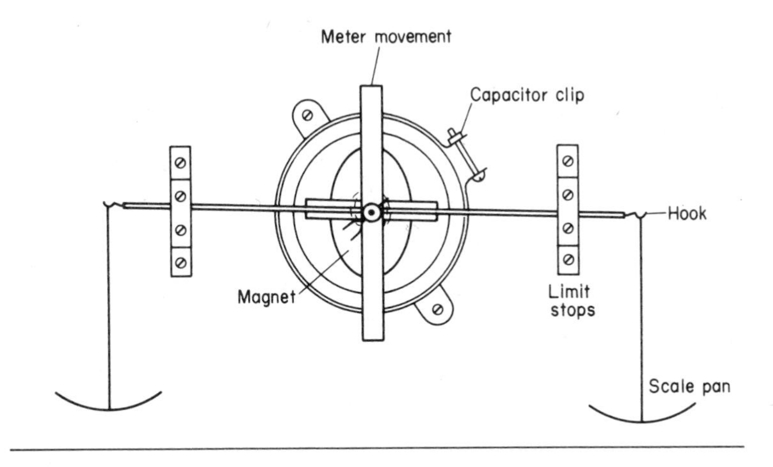

Fig. 52. Layout of the microbalance.

variety (the magnet is inside the coil) but meters of more traditional design with external magnets, could be used, although they are considerably large and heavier. The layout of the balance is shown in Fig. 52. The pointer an the outer part of the counter-weight are carefully removed from the meter and replaced by two lengths of stainless-steel hypodermic needle, taken fron 19 G × 1½ (40 mm 11/10) needles, which are fixed to the meter movemen by epoxy resin adhesive. The outer ends of the tubing arms are fitted wit hooks made from 22 swg copper wire which fits inside the tubing and ar secured with epoxy adhesive. The scale pans are made from thin brass shin and hung on the hooks at the ends of the beam by lengths of very fine coppe wire. In order to reduce the risk of damage to the movement during handling the beam is fitted at each side with limit stops (Fig. 53). The hooks (Fig. 54 are shaped so that the point at which the scale pans pivot is exactly in line wit the central pivot of the movement.

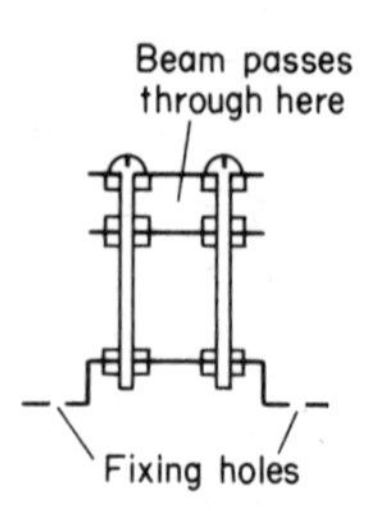

Fig. 53. Detail of the limit stops.

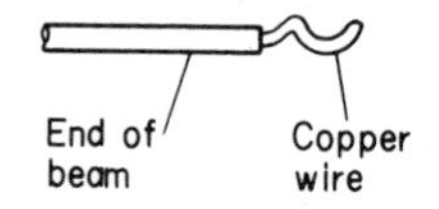

Fig. 54. Detail of the ends of the beam.

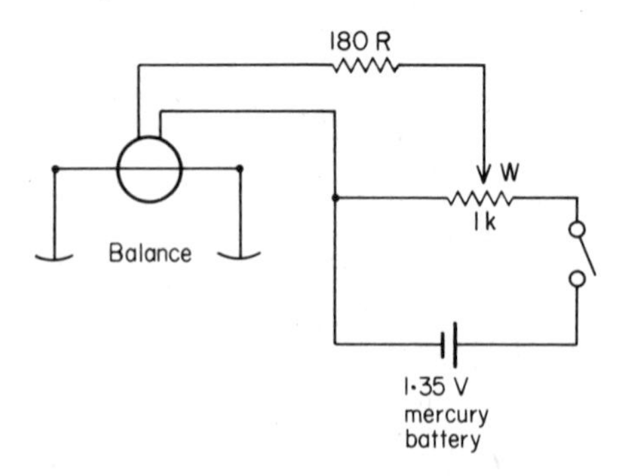

Fig. 55. Circuit of the simple microbalance.

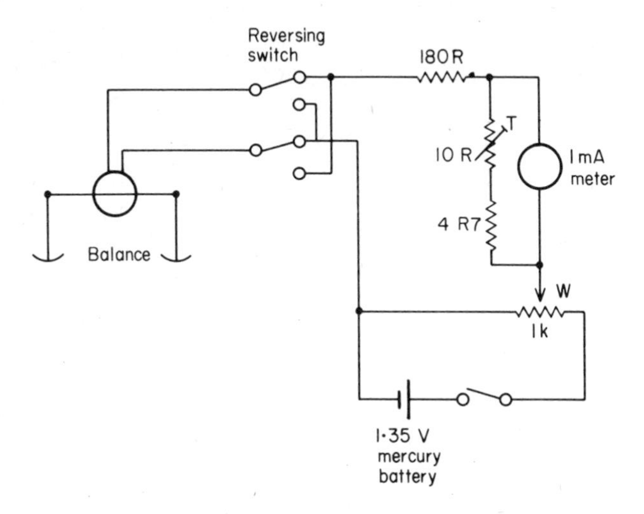

Fig. 56. Circuit of the microbalance with meter indication.

A very simple circuit for the microbalance is given in Fig. 55. The potentiometer (W) selects the amount of current which is fed through the meter movement in order to centre the beam, and since the force produced on the beam is proportional to the current flowing through the coil of the movement, the potentiometer can be calibrated directly in milligrams. The circuit is arranged so that the balance has a range of 50 mg indicated on the potentiometer. Larger weights can, of course, be measured by adding weights to the scale pans. Figure 56 shows a more elaborate circuit which uses a second moving-coil meter to indicate the weight, and also incorporates a reversing switch in the connections to the balance movement. The meter scale is linear and can be calibrated to a range of ± 50 mg by adjusting the preset potentiometer (T).

Moving-coil meters are very well constructed and have jewelled bearings, which are spring-loaded to reduce the chance of damage. Provided the microbalance is carefully made, it can be accurate to about 1 mg.

IV. Artificial Lighting

It is generally desirable that artificial lighting in laboratory simulations of the real world should have a spectral output similar to sunlight. Unfortunately, there is no lamp which does this, and either tungsten lighting or some type of fluorescent lamp generally has to be used. The characteristics of a "daylight"

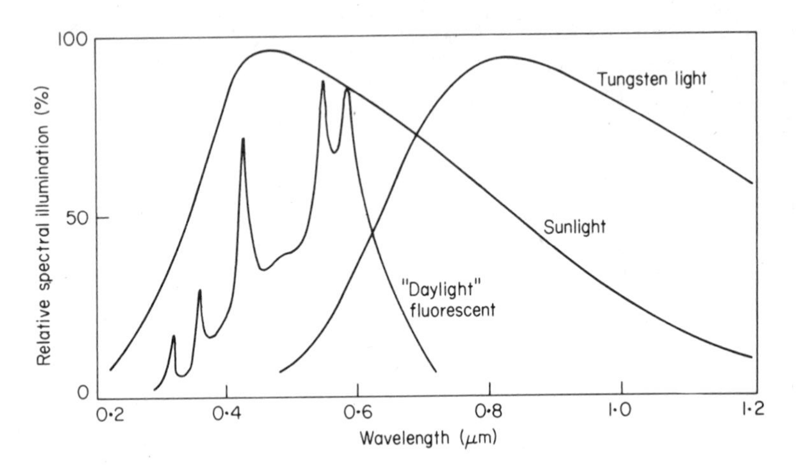

Fig. 57. The output of a "daylight" flourescent tube compared with sunlight and tungsten light.

tube are shown in Fig. 57, compared with sunlight and a tungsten lamp, and it will be seen that while this tube seems to produce white light to the human eye, it is very far from being a good simulation of sunlight; in particular, the longer wavelengths are almost completely absent. Tungsten light is particularly deficient at the shorter wavelengths, and a combination of both fluorescent and tungsten lighting has the advantage that even if the spectral distribution is not a good mimic of sunlight, at least most of the wavelengths will be present.

There are a number of special lamps produced for plant propagation which have the spectral lines that are most important for photosynthesis emphasised. While these lamps do appear to be effective when fitted in greenhouses, there is no guarantee that they will be equally effective in, for example, an insect culture, and this limitation should be taken into consideration when studying pests in a greenhouse.

All lamps get hot, and can interfere with temperature control if they are fitted inside an environmental cabinet. If possible, it is better to locate them outside such a cabinet, and use a glass window (which will stop the long-wave radiation) to transmit the light to the inside. If this is not possible, some of the effects of heat can be minimised by using fluorescent lighting with just the tube inside the cabinet, and locating the choke unit outside.

All lamps run from alternating-current mains will flicker, and since the lamps light on both half-cycles, they flicker at twice the mains frequency. This means a frequency of 100 Hz in most parts of the world, or 120 Hz in the United States. Miall (1978) has shown that these frequencies are below the

icker fusion frequencies of some insects and the flicker frequency of light-
g in insect cultures has been shown to affect insect behaviour (Van Praagh,
975). It is difficult to know how important these effects are, but laboratory-
ultured insects have the reputation of being rather poor specimens when
ompared with wild insects or with even those reared by amateur entomo-
gists in their garden sheds. Fortunately, there is a way out of the flicker
roblem, and this involves the use of low-voltage inverter fluorescent lamps
overed by direct-current supplies. Inverter fluorescents normally have
iternal oscillators operating at very high frequencies (often 1000 Hz), well
bove insect flicker fusion frequencies. They can be powered from
ickel – cadmium batteries, which are kept continuously charged.

V. Automatic Recording and Data Logging

utomatic recording is one of the strongest temptations in microclimate
neasurement and although it can be a very convenient way of collecting data,
here are occasions where the temptation should be resisted. The simple
pproach to microclimate recording involves the experimenter going out into
he field with a set of portable instruments, measuring a wide range of
actors, and writing the measurements in a notebook. There is a good deal to
e said in favour of this approach, and it should always be the first stage in
ny investigation: automatic recording is a great convenience once you know
vhat you need to measure. The notebook method produces a rather infre-
uent record of a large number of measurements, while a pen recorder pro-
luces a continuous record of a limited number of measurements. Digital data
ogging, where the analogue input signal is turned into a number and printed
n a roll of paper (or recorded on magnetic tape) and where a large number of
nputs can be scanned in sequence, gives us a discontinuous but frequent
ecord of a large number of measurements. Digital data logging lends itself
ery well to automatic data processing and will almost certainly replace conti-
uous recording in the majority of applications. However, the notebook will
emain the cheapest, lightest and most versatile system.

Most of the circuits included in this book can be adapted for use with auto-
natic recording systems. The exceptions are instruments like the thermo-
ouple meter, where adjustments have to be made for changes in reference
emperature. 100 μA meters usually have an internal resistance of 1300 Ω,
which means that the voltage developed across them at full-scale deflection
will be 130 mV. Meters of 1 mA usually have a resistance of 75 Ω, which will
result in a voltage of 75 mV being developed across them at full deflection. In
many of the circuits in this book, a 100 mA meter will be found to be
connected in series with a resistance of 10k (or a potentiometer which will in

fact be adjusted to this value). Similarly, in many cases, a 1 mA meter will be found to be connected in series with a 1k resistor. In both cases, the voltage developed across the meter and resistance in series will be 1 V at full deflection of the meter.

By far the most convenient type of sensor for use with recorders or data logging systems are those which indicate the factor being monitored by a change in resistance. Thus, photoconductive cells, thermistors and resistive humidity sensors are particularly useful. Silicon solar cell solarimeters are also very convenient since they generate a large output.

6
Humidity Calculations and Tables

The humidity calculations covered in this chapter fall into two groups: the first group is concerned with expressing the output of psychrometers and dew-point equipment in terms of relative humidity, and the second is concerned with deriving other microclimate parameters (saturation deficit, vapour pressure and absolute humidity) from relative humidity and air temperature.

A. Saturation Vapour Pressure

In order to calculate relative humidity from the temperature readings obtained from psychrometers and dew-point equipment, we must first know the saturation vapour pressure at these temperatures. The vapour pressure of saturated air can be related to its temperature by the following empirical equation, the form of which was suggested to me by my colleague, Dr K. E. Machin:

$$\log_{10} e_{s(t)} = 9.24349 - \frac{2305}{t} - \frac{500}{t^2} - \frac{100\,000}{t^3}$$

where $e_{s(t)}$ is the saturation vapour pressure at temperature t, in millibars; t is the temperature in degrees Kelvin (°C + 273). This formula gives a value to within ± 0.1 mbar of those given by Monteith (1973) over the range − 5 to + 34°C.

The relationship between saturation vapour pressure and temperature is also given in Table 4 (data from Monteith, 1973).

B. Psychrometers

The vapour pressure can be calculated from the dry and wet bulb temperatures

Table 4

Saturation Vapour Pressure versus Temperature (°C).

T (°C)	$e_{s(t)}$ (m bar)	T (°C)	$e_{s(t)}$ (m bar)
−5·0	4·21	20·5	24·12
−4·5	4·38	21·0	24·86
−4·0	4·55	21·5	25·65
−3·5	4·73	22·0	26·43
−3·0	4·90	22·5	27·26
−2·5	5·09	23·0	28·09
−2·0	5·25	23·5	28·96
−1·5	5·48	24·0	29·83
−1·0	5·68	24·5	30·75
−0·5	5·90	25·0	31·67
0·0	6·11	25·5	32·64
		26·0	33·61
0·5	6·34	26·5	34·63
1·0	6·57	27·0	35·65
1·5	6·81	27·5	36·73
2·0	7·05	28·0	37·80
2·5	7·32	28·5	38·93
3·0	7·58	29·0	40·06
3·5	7·86	29·5	41·25
4·0	8·13	30·0	42·43
4·5	8·43		
5·0	8·72	30·5	43·68
5·5	9·04	31·0	44·93
6·0	9·35	31·5	46·24
6·5	9·68	32·0	47·55
7·0	10·01	32·5	48·93
7·5	10·37	33·0	50·31
8·0	10·72	33·5	51·76
8·5	11·10	34·0	53·20
9·0	11·47	34·5	54·72
9·5	11·87	35·0	56·24
10·0	12·27	35·5	57·83
		36·0	59·42
10·5	12·70	36·5	61·09
11·0	13·12	37·0	62·76
11·5	13·57	37·5	64·51
12·0	14·02	38·0	66·26
12·5	14·50	38·5	68·10
13·0	14·97	39·0	69·93
13·5	15·48	39·5	71·86
14·0	15·98	40·0	73·78
14·5	16·51		
15·0	17·04	40·5	75·79

able 4 (*Continued*)

T (°C)	$e_{s(t)}$ (m bar)	T (°C)	$e_{s(t)}$ (m bar)
15·5	17·61	41·0	77·80
16·0	18·17	41·5	79·91
16·5	18·77	42·0	82·02
17·0	19·37	42·5	84·22
17·5	20·00	43·0	86·42
18·0	20·63	43·5	88·73
18·5	21·30	44·0	91·03
19·0	21·96	44·5	93·45
19·5	22·67	45·0	95·86
20·0	23·37		

f a psychrometer by the following formula:

$$e = e_{s(w)} - A(t_d - t_w)$$

here e is the vapour pressure in millibars, $e_{s(w)}$ is the saturation vapour pres-ure at the temperature of the wet bulb, also in millibars, t_d is the dry bulb emperature and t_w the wet bulb temperature, both in degrees Celsius, A is a onstant which depends on the rate of aspiration and the barometric pres-ure. For practical purposes, the barometric pressure complication can be gnored below 1000 metres. At higher altitudes, the value of A should be nultiplied by the pressure in millibars and divided by 1000.

Non-aspirated psychrometers are usually assumed to have a natural aspira-ion of about 1 m/s and the value of A taken as 0.799. When the temperature f the wet bulb has fallen below zero, ice will form and the value for A is then aken as 0.720 (Met. Office, 1964a). The formula for a non-aspirated sychrometer thus becomes:

$$e = e_{s(w)} - 0.799\,(t_d - t_w)$$

or normal conditions, and

$$e = e_{s(w)} - 0.720\,(t_d - t_w)$$

here ice has formed on the wet bulb.

Aspirated psychrometers are assumed to have a rate of aspiration of at east 3 m/s and the value of A is taken as 0.666 under normal conditions and .594 when ice has formed on the wet bulb (Met. Office, 1964b). The formula or an aspirated psychrometer thus becomes:

$$e = e_{s(w)} - 0.666\ (t_d - t_w)$$

for normal conditions, and

$$e = e_{s(w)} - 0.594\ (t_d - t_w)$$

when ice has formed on the wet bulb.

Relative humidity may be calculated from the formula:

$$\text{RH}\ (\%) = \frac{e}{e_{s(t)}} \times 100$$

where e is the vapour pressure and $e_{s(t)}$ is the saturation vapour pressure at the dry bulb temperature.

It should be noted that the saturation vapour pressure used earlier in the calculation of the vapour pressure was at the wet bulb temperature. The saturation vapour pressure above is at the dry bulb temperature. The relationship between wet and dry bulb temperatures and relative humidity for non-aspirated and aspirated psychrometers is given in Tables 5 and 6.

C. Dew-point Equipment

Relative humidity may be calculated from dew-point and air temperature as follows:

$$\text{RH} = \frac{e_{s(d)} \times 100}{e_{s(t)}}$$

where $e_{s(d)}$ is the saturation vapour pressure at the dew-point and $e_{s(t)}$ is the saturation vapour pressure at air temperature.

The relationship between dew-point, air temperature and relative humidity is given in Table 7.

D. Choice of a Humidity Parameter

The most commonly used parameter for expressing atmospheric water content is relative humidity, which is the amount of water in the air divided by the amount in saturated air at the same temperature. The degree of saturation can also be expressed as the saturation deficit, which is the difference between the amount of water in the air and the amount in saturated air at the same temperature. These amounts of water in the air are usually described in terms of vapour pressures, and the saturation deficit is also called the vapour pressure deficit. Relative humidity and vapour pressure deficit are both ways of describing the degree of saturation of the air, either as a ratio or as a difference. In systems which equilibrate to the humidity of the air, such as the

bulb (°C)	0·2	0·4	0·6	0·8	1·0	1·2	1·4	1·6	1·8	2·0	2·5	3·0	3·5	4·0	4·5	5·0	6·0	7·0	8·0	9·0	10	11	12	13	14	15	16
0	96	92	88	84	80	76	73	69	65	61	52	43	33	24	16	7											
1	96	92	88	85	81	78	75	71	67	64	55	46	37	29	20	12											
2	96	93	89	85	82	78	75	71	67	64	58	49	41	33	24	16	1										
3	96	93	89	86	83	79	76	72	69	66	57	49	44	36	28	21	6										
4	97	93	90	87	83	80	77	74	70	67	59	51	44	36	32	25	10										
5	97	94	90	87	84	81	78	75	72	69	61	53	46	39	31	24	15	2									
6	97	94	91	88	85	82	79	76	73	70	63	55	48	41	34	28	14	6									
7	97	94	91	88	85	82	80	77	74	71	64	57	50	44	37	31	18	5									
8	97	94	92	89	86	83	80	78	75	72	66	59	52	46	40	33	21	9									
9	97	95	92	89	86	84	81	79	76	73	67	61	54	48	42	36	24	13	2								
10	97	95	92	90	87	84	82	79	77	74	68	62	56	50	44	38	27	16	5								
11	97	95	92	90	87	85	83	80	78	75	69	63	58	52	46	41	30	19	9								
12	98	95	93	90	88	86	83	81	78	76	70	65	59	54	48	43	32	22	12	3							
13	98	95	93	91	88	86	84	81	79	77	71	66	61	55	50	45	35	26	17	6							
14	98	95	93	91	89	86	84	82	80	78	72	67	62	57	52	47	37	28	18	10	1						
15	98	96	93	91	89	87	85	83	81	78	73	68	63	58	53	49	39	30	21	13	4						
16	98	96	94	91	89	87	85	83	81	79	74	69	64	60	55	50	41	32	24	16	7						
17	98	96	94	92	90	88	86	84	82	80	75	70	65	61	56	52	43	34	26	18	10	3					
18	98	96	94	92	90	88	86	84	82	80	76	71	66	62	58	53	45	36	28	21	13	6					
19	98	96	94	92	90	88	86	85	83	81	76	72	67	63	59	55	46	38	31	23	16	9					
20	98	96	94	92	91	89	87	85	83	81	77	73	68	64	60	56	48	40	33	25	18	11	5				
21	98	96	94	93	91	89	87	85	84	82	78	73	69	65	61	57	49	42	34	27	21	14	8	1			
22	98	96	95	93	91	89	88	86	84	82	78	74	70	66	62	58	51	43	36	29	23	16	10	4			
23	98	96	95	93	91	90	88	86	84	83	79	75	71	67	63	59	52	45	38	31	25	19	13	7	1		
24	98	97	95	93	91	90	88	86	85	83	79	75	71	68	64	60	53	46	39	33	27	21	15	9	4		
25	98	97	95	93	92	90	88	87	85	84	80	76	72	68	65	61	54	47	41	35	29	23	17	11	6	1	
26	98	97	95	93	92	90	89	87	86	84	80	76	73	69	66	62	55	49	42	36	30	25	19	14	8	3	
27	98	97	95	94	92	90	89	87	86	84	81	77	73	70	66	63	56	50	44	38	32	26	21	16	11	6	1
28	98	97	95	94	92	91	89	88	86	85	81	77	74	70	67	64	57	51	45	39	33	28	23	18	13	8	3
29	98	97	95	94	92	91	89	88	86	85	81	78	74	71	68	65	58	52	46	40	35	30	24	19	15	10	5
30	98	97	95	94	93	91	90	88	87	85	82	78	75	72	68	65	59	53	47	42	36	31	26	21	16	12	8
31	99	97	96	94	93	91	90	88	87	86	82	79	75	72	69	66	60	54	48	43	38	32	28	23	18	14	9
32	99	97	96	94	93	91	90	89	87	86	83	79	76	73	70	67	61	55	49	44	39	34	29	24	20	16	11
33	99	97	96	94	93	91	90	89	87	86	83	80	76	73	70	67	61	56	50	45	40	35	30	26	21	17	13
34	99	97	96	94	93	92	90	89	87	86	83	80	77	74	71	68	62	56	51	46	41	36	32	27	23	19	15
35	99	98	97	95	93	92	90	90	88	87	83	80	77	74	71	68	63	57	52	47	42	37	33	29	24	20	16
36	99	98	97	95	93	92	90	90	88	87	84	81	78	75	72	69	63	58	53	48	43	39	34	30	26	22	18
37	99	98	97	95	93	92	90	90	88	87	84	81	78	75	72	69	64	59	54	49	44	40	35	31	27	23	19
38	99	98	97	95	93	92	90	90	88	87	84	81	78	75	73	70	65	59	54	50	45	41	36	32	28	24	21
39	99	98	97	95	94	93	91	91	89	87	84	81	78	75	73	70	65	60	55	50	46	41	37	33	29	26	22

Table 6. Relative Humidity: Aspirated Psychrometer

Dry bulb	Wet bulb depression (°C)																									
(°C)	0·2	0·4	0·6	0·8	1·0	1·2	1·4	1·6	1·8	2·0	2·5	3·0	3·5	4·0	4·5	5·0	6·0	7·0	8·0	9·0	10	11	12	13	14	15	16
0	96	93	89	86	82	79	75	72	69	65	57	49	41	33	25	17	2										
1	97	93	90	86	83	81	77	74	71	68	60	52	44	36	29	21	7										
2	97	93	90	87	84	80	77	74	71	68	62	54	47	40	32	25	11										
3	97	94	91	88	84	81	78	75	72	69	62	54	50	43	36	29	16	3									
4	97	94	91	88	85	82	79	76	73	70	63	56	49	42	39	33	20	7									
5	97	94	91	88	86	83	80	77	74	72	65	58	51	45	38	32	25	12									
6	97	94	92	89	86	83	81	78	75	73	66	60	53	47	41	35	23	16	5								
7	97	95	92	89	87	84	81	79	76	74	67	61	55	49	43	37	26	14	9								
8	97	95	92	90	87	85	82	80	77	75	69	63	57	51	45	40	29	18	7	3							
9	97	95	93	90	88	85	83	80	78	76	70	64	58	53	47	42	31	21	11	1							
10	98	95	93	90	88	86	83	81	79	77	71	65	60	54	49	44	34	24	14	5							
11	98	95	93	91	88	86	84	82	80	77	72	66	61	56	51	46	36	26	17	8							
12	98	95	93	91	89	87	84	82	80	78	73	68	63	57	53	48	38	29	20	11	3						
13	98	96	93	91	89	87	85	83	81	79	74	69	64	59	54	49	40	31	23	14	6						
14	98	96	94	92	90	87	85	83	81	79	74	70	65	60	56	51	42	33	25	17	9	2					
15	98	96	94	92	90	88	86	84	82	80	75	71	66	61	57	52	44	36	27	20	12	5					
16	98	96	94	92	90	88	86	84	82	81	76	71	67	62	58	54	46	37	30	22	15	8	1				
17	98	96	94	92	90	89	87	85	83	81	77	72	68	64	59	55	47	39	32	24	17	10	4				
18	98	96	94	92	91	89	87	85	83	82	77	73	69	65	60	56	49	41	34	27	20	13	7				
19	98	96	94	93	91	89	87	86	84	82	78	74	70	65	61	58	50	43	35	29	22	15	9	3			
20	98	96	95	93	91	89	88	86	84	83	78	74	70	66	62	59	51	44	37	30	24	18	12	6			
21	98	96	95	93	91	90	88	86	85	83	79	75	71	67	63	60	52	46	39	32	26	20	14	8	3		
22	98	97	95	93	92	90	88	87	85	83	79	76	72	68	64	61	54	47	40	34	28	22	16	11	5		
23	98	97	95	93	92	90	89	87	85	84	80	76	72	69	65	62	55	48	42	36	30	24	18	13	8	3	
24	98	97	95	94	92	90	89	87	86	84	80	77	73	69	66	62	56	49	43	37	31	26	20	15	10	5	
25	98	97	95	94	92	91	89	87	86	84	81	77	74	70	67	63	57	50	44	38	33	27	22	17	12	7	3
26	98	97	95	94	92	91	89	88	86	85	81	78	74	71	67	64	58	51	46	40	34	29	24	19	14	9	5
27	98	97	95	94	92	91	89	88	87	85	82	78	75	71	68	65	59	52	47	41	36	30	25	21	16	11	7
28	98	97	95	94	93	91	90	88	87	85	82	79	75	72	69	65	59	53	48	42	37	32	27	22	18	13	9
29	99	97	96	94	93	91	90	88	87	86	82	79	76	72	69	66	60	54	49	43	38	33	28	24	19	15	11
30	99	97	96	94	93	91	90	89	87	86	83	79	76	73	70	67	61	55	50	44	39	34	30	25	21	17	13
31	99	97	96	94	93	92	90	89	88	86	83	80	77	73	70	67	62	56	51	45	41	36	31	27	22	18	14
32	99	97	96	94	93	92	90	89	88	86	83	80	77	74	71	68	62	57	52	46	42	37	32	28	24	20	16
33	99	97	96	94	93	92	90	89	88	87	83	80	77	74	71	69	63	58	52	47	43	38	34	29	25	21	17
34	99	98	97	95	93	92	90	89	88	87	84	81	78	75	72	69	64	58	53	48	44	39	35	30	26	23	19
35	99	98	97	95	93	92	90	89	88	87	84	81	78	75	72	70	64	59	54	49	44	40	36	32	28	24	20
36	99	98	97	95	94	92	90	89	88	87	84	81	78	76	73	70	65	60	55	50	45	41	37	33	29	25	21

Table 7

Relative Humidity: Dew-point and Air Temperature

Dew-point depression (°C)	Air temperature (°C)						
	0	5	10	15	20	25	30
0·5	96·5	96·6	96·7	96·8	96·9	97·0	97·1
1·0	93·1	93·3	93·5	93·8	94·0	94·2	94·3
1·5	89·8	90·1	90·5	90·8	91·1	91·4	91·6
2·0	86·6	87·0	87·5	87·9	88·3	88·6	89·0
2·5	83·5	84·0	84·6	85·1	85·5	86·0	86·4
3·0	80·5	81·1	81·7	82·3	82·9	83·4	83·9
3·5	77·6	78·3	79·0	79·7	80·3	80·9	81·4
4·0	74·8	75·6	76·4	77·1	77·8	78·4	79·1
4·5	72·1	73·0	73·8	74·6	75·3	76·0	76·7
5·0	69·5	70·4	71·3	72·1	72·9	73·7	74·5
5·5	67·0	67·9	68·9	69·8	70·6	71·5	72·3
6·0	64·5	65·5	66·5	67·5	68·4	69·3	70·1
6·5	62·1	63·2	64·3	65·3	66·2	67·1	68·0
7·0	59·8	61·0	62·1	63·1	64·1	65·1	66·0
7·5	57·6	58·8	59·9	61·0	62·0	63·1	64·0
8·0	55·5	56·7	57·8	59·0	60·0	61·1	62·1
8·5	53·4	54·7	55·8	57·0	58·1	59·2	60·2
9·0	51·4	52·7	53·9	55·1	56·2	57·3	58·4
9·5	49·5	50·8	52·0	53·2	54·4	55·5	56·6
10	47·6	48·9	50·2	51·4	52·6	53·8	54·9
11		45·4	46·7	48·0	49·2	50·4	51·6
12		42·1	43·5	44·8	46·0	47·3	48·5
13		39·1	40·4	41·7	43·0	44·3	45·5
14		36·2	37·6	38·9	40·2	41·5	42·7
15		33·5	34·9	36·2	37·5	38·8	40·1
16			32·4	33·7	35·0	36·3	37·6
17			30·0	31·4	32·7	33·9	35·2
18			27·9	29·2	30·4	31·7	33·0
19			25·8	27·1	28·4	29·6	30·9
20			23·9	25·2	26·4	27·7	28·9
21				23·4	24·6	25·8	27·0
22				21·7	22·9	24·1	25·3
23				20·1	21·3	22·4	23·6
24				18·6	19·8	20·9	22·1
25				17·2	18·4	19·5	20·6

water content of porous materials or the nectar of flowers (Corbet *et al.* 1979b), relative humidity is more appropriate since the relationship betwee the water content of materials and solutions at equilibrium is nearly indepen dent of temperature. Vapour pressure deficit is more appropriate fo dynamic situations which are far from being at equilibrium, since its relation ship to the rate of evaporation of water is independent of temperature. It ha been shown empirically that in some biological studies, better correlation are obtained between biological events and humidity if the latter is expresse in vapour pressure deficit rather than relative humidity (Williams an Brochu, 1969). However, most biological systems involve aqueous solution which are not equilibrated with the air, but are sufficiently concentrated tha they cannot be treated as pure water, and the choice of a humidity paramete may not be clear. In cases of doubt, it is probably better to record the relativ humidity, but also noting the air temperature so that the vapour pressur deficit can be calculated should this be required.

The amount of water in the air can be expressed in ways that take n account of the degree of saturation, such as vapour pressure and absolut humidity. Vapour pressure is not generally useful as a humidity parameter but is used in the calculation of other parameters. Absolute humidity, whic is the weight of water per unit volume of air, is much more useful. If it i plotted against distance or time, it can tell us where and when the processes o evaporation and condensation are taking place (Unwin, 1978).

E. Vapour Pressure

The vapour pressure (Table 8) can be calculated using the formula:

$$e = \frac{e_{s(t)} \times \mathrm{RH}}{100}$$

where e is the vapour pressure in millibars, $e_{s(t)}$ is the saturation vapou pressure at air temperature, also in millibars, and RH is the relative humidit in %.

F. Vapour Pressure Deficit

The vapour pressure deficit (or saturation deficit) (Table 9) can be calculate as follows:

$$\mathrm{VPD} = e_{s(t)} - e$$

where VPD is the vapour pressure deficit in millibars.

Temp. (°C)	Relative humidity (%) 100	96	92	88	84	80	76	72	68	64	60	56	52	48	44	40	36	32	28	24	20
0	6·1	5·9	5·6	5·4	5·1	4·9	4·6	4·4	4·2	3·9	3·7	3·4	3·2	2·9	2·7	2·4	2·2	2·0	1·7	1·5	1·2
1	6·6	6·3	6·0	5·8	5·5	5·3	5·0	4·7	4·5	4·2	3·9	3·7	3·4	3·2	2·9	2·6	2·4	2·1	1·8	1·6	1·3
2	7·1	6·8	6·5	6·2	5·9	5·6	5·4	5·1	4·8	4·5	4·2	3·9	3·7	3·4	3·1	2·8	2·5	2·3	2·0	1·7	1·4
3	7·6	7·3	7·0	6·7	6·4	6·1	5·8	5·5	5·2	4·9	4·5	4·2	3·9	3·6	3·3	3·0	2·7	2·4	2·1	1·8	1·5
4	8·1	7·8	7·5	7·2	6·8	6·5	6·2	5·9	5·5	5·2	4·9	4·6	4·2	3·9	3·6	3·3	2·9	2·6	2·3	2·0	1·6
5	8·7	8·4	8·0	7·7	7·3	7·0	6·6	6·3	5·9	5·6	5·2	4·9	4·5	4·2	3·8	3·5	3·1	2·8	2·4	2·1	1·7
6	9·4	9·0	8·6	8·2	7·9	7·5	7·1	6·7	6·4	6·0	5·6	5·2	4·9	4·5	4·1	3·7	3·4	3·0	2·6	2·2	1·9
7	10·0	9·6	9·2	8·8	8·4	8·0	7·6	7·2	6·8	6·4	6·0	5·6	5·2	4·8	4·4	4·0	3·6	3·2	2·8	2·4	2·0
8	10·7	10·3	9·9	9·4	9·0	8·6	8·1	7·7	7·3	6·9	6·4	6·0	5·6	5·1	4·7	4·3	3·9	3·4	3·0	2·6	2·1
9	11·5	11·0	10·6	10·1	9·6	9·2	8·7	8·3	7·8	7·3	6·9	6·4	6·0	5·5	5·0	4·6	4·1	3·7	3·2	2·8	2·3
10	12·3	11·8	11·3	10·8	10·3	9·8	9·3	8·8	8·3	7·9	7·4	6·9	6·4	5·9	5·4	4·9	4·4	3·9	3·4	2·9	2·5
11	13·1	12·6	12·1	11·5	11·0	10·5	10·0	9·4	8·9	8·4	7·9	7·3	6·8	6·3	5·8	5·2	4·7	4·2	3·7	3·1	2·6
12	14·0	13·5	12·9	12·3	11·8	11·2	10·7	10·1	9·5	9·0	8·4	7·9	7·3	6·7	6·2	5·6	5·0	4·5	3·9	3·4	2·8
13	15·0	14·4	13·8	13·2	12·6	12·0	11·4	10·8	10·2	9·6	9·0	8·4	7·8	7·2	6·6	6·0	5·4	4·8	4·2	3·6	3·0
14	16·0	15·3	14·7	14·1	13·4	12·8	12·1	11·5	10·9	10·2	9·6	8·9	8·3	7·7	7·0	6·4	5·8	5·1	4·5	3·8	3·2
15	17·0	16·4	15·7	15·0	14·3	13·6	13·0	12·3	11·6	10·9	10·2	9·5	8·9	8·2	7·5	6·8	6·1	5·5	4·8	4·1	3·4
16	18·2	17·4	16·7	16·0	15·3	14·5	13·8	13·1	12·4	11·6	10·9	10·2	9·4	8·7	8·0	7·3	6·5	5·8	5·1	4·4	3·6
17	19·4	18·6	17·8	17·0	16·3	15·5	14·7	13·9	13·2	12·4	11·6	10·8	10·1	9·3	8·5	7·7	7·0	6·2	5·4	4·6	3·9
18	20·6	19·8	19·0	18·2	17·3	16·5	15·7	14·9	14·0	13·2	12·4	11·6	10·7	9·9	9·1	8·3	7·4	6·6	5·8	5·0	4·1
19	22·0	21·1	20·2	19·3	18·4	17·6	16·7	15·8	14·9	14·1	13·2	12·3	11·4	10·5	9·7	8·8	7·9	7·0	6·1	5·3	4·4
20	23·4	22·4	21·5	20·6	19·6	18·7	17·8	16·8	15·9	15·0	14·0	13·1	12·2	11·2	10·3	9·3	8·4	7·5	6·5	5·6	4·7
21	24·9	23·9	22·9	21·9	20·9	19·9	18·9	17·9	16·9	15·9	14·9	13·9	12·9	11·9	10·9	9·9	8·9	8·0	7·0	6·0	5·0
22	26·4	25·4	24·3	23·3	22·2	21·1	20·1	19·0	18·0	16·9	15·9	14·8	13·7	12·7	11·6	10·6	9·5	8·5	7·4	6·3	5·3
23	28·1	27·0	25·8	24·7	23·6	22·5	21·3	20·2	19·1	18·0	16·9	15·7	14·6	13·5	12·4	11·2	10·1	9·0	7·9	6·7	5·6
24	29·8	28·6	27·4	26·3	25·1	23·9	22·7	21·5	20·3	19·9	17·9	16·7	15·5	14·3	13·1	11·9	10·7	9·5	8·4	7·2	6·0
25	31·7	30·4	29·1	27·9	26·6	25·3	24·1	22·8	21·5	20·3	19·0	17·7	16·5	15·2	13·9	12·7	11·4	10·1	8·9	7·6	6·3
26	33·6	32·3	30·9	29·6	28·2	26·9	25·5	24·2	22·9	21·5	20·2	18·8	17·5	16·1	14·8	13·4	12·1	10·8	9·4	8·1	6·7
27	35·7	34·2	32·8	31·4	29·9	28·5	27·1	25·7	24·2	22·8	21·4	20·0	18·5	17·1	15·7	14·3	12·8	11·4	10·0	8·6	7·1
28	37·8	36·3	34·8	33·3	31·8	30·2	28·7	27·2	25·7	24·2	22·7	21·2	19·7	18·1	16·6	15·1	13·6	12·1	10·6	9·1	7·6
29	40·1	38·5	36·9	35·3	33·7	32·0	30·4	28·8	27·2	25·6	24·0	22·4	20·8	19·2	17·6	16·0	14·4	12·8	11·2	9·6	8·0
30	42·4	40·7	39·0	37·3	35·6	33·9	32·2	30·5	28·9	27·2	25·5	23·8	22·1	20·4	18·7	17·0	15·3	13·6	11·9	10·2	8·5
31	44·9	43·1	41·3	39·5	37·7	35·9	34·1	32·3	30·6	28·8	27·0	25·2	23·4	21·6	19·8	18·0	16·2	14·4	12·6	10·8	9·0
32	47·6	45·6	43·7	41·8	39·9	38·0	36·1	34·2	32·3	30·4	28·5	26·6	24·7	22·8	20·9	19·0	17·1	15·2	13·3	11·4	9·5
33	50·3	48·3	46·3	44·3	42·3	40·2	38·2	36·2	34·2	32·2	30·2	28·2	26·2	24·1	22·1	20·1	18·1	16·1	14·1	12·1	10·1
34	53·2	51·1	48·9	46·8	44·7	42·6	40·4	38·3	36·2	34·0	31·9	29·8	27·7	25·5	23·4	21·3	19·2	17·0	14·9	12·8	10·6
35	56·2	54·0	51·7	49·5	47·2	45·0	42·7	40·5	38·2	36·0	33·7	31·5	29·2	27·0	24·7	22·5	20·2	18·0	15·7	13·5	11·2
36	59·4	57·0	54·7	52·3	49·9	47·5	45·2	42·8	40·4	38·0	35·7	33·3	30·9	28·5	26·1	23·8	21·4	19·0	16·6	14·3	11·9
37	62·8	60·2	57·7	55·2	52·7	50·2	47·7	45·2	42·7	40·2	37·7	35·1	32·6	30·1	27·6	25·1	22·6	20·1	17·6	15·1	12·6
38	66·3	63·6	61·0	58·3	55·7	53·0	50·4	47·7	45·1	42·4	39·8	37·1	34·5	31·8	29·2	26·5	23·9	21·2	18·6	15·9	13·3
39	69·9	67·1	64·3	61·5	58·7	55·9	53·1	50·3	47·6	44·8	42·0	39·2	36·4	33·6	30·8	28·0	25·2	22·4	19·6	16·8	14·0

G. Absolute Humidity

The absolute humidity (Table 10) is given by

$$AH = \frac{216.7 \times e}{t + 273}$$

where AH is the absolute humidity in grams per cubic metre of air, and t is the air temperature in °C.

Table 10. Absolute Humidity (in grams per cubic metre) from Relative Humidity and Temperature

Temp.	Relative humidity (%)																				
(°C)	100	96	92	88	84	80	76	72	68	64	60	56	52	48	44	40	36	32	28	24	20
0	4·8	4·7	4·5	4·3	4·1	3·9	3·7	3·5	3·3	3·1	2·9	2·7	2·5	2·3	2·1	1·9	1·7	1·6	1·4	1·2	1·0
1	5·2	5·0	4·8	4·6	4·4	4·2	3·9	3·7	3·5	3·3	3·1	2·9	2·7	2·5	2·3	2·1	1·9	1·7	1·5	1·2	1·0
2	5·6	5·3	5·1	4·9	4·7	4·4	4·2	4·0	3·8	3·6	3·3	3·1	2·9	2·7	2·4	2·2	2·0	1·8	1·6	1·3	1·1
3	6·0	5·7	5·5	5·2	5·0	4·8	4·5	4·3	4·0	3·8	3·6	3·3	3·1	2·9	2·6	2·4	2·1	1·9	1·7	1·4	1·2
4	6·4	6·1	5·9	5·6	5·3	5·1	4·8	4·6	4·3	4·1	3·8	3·6	3·3	3·1	2·8	2·5	2·3	2·0	1·8	1·5	1·3
5	6·8	6·5	6·3	6·0	5·7	5·4	5·2	4·9	4·6	4·4	4·1	3·8	3·5	3·3	3·0	2·7	2·4	2·2	1·9	1·6	1·4
6	7·3	7·0	6·7	6·4	6·1	5·8	5·5	5·2	4·9	4·6	4·4	4·1	3·8	3·5	3·2	2·9	2·6	2·3	2·0	1·7	1·5
7	7·7	7·4	7·1	6·8	6·5	6·2	5·9	5·6	5·3	5·0	4·6	4·3	4·0	3·7	3·4	3·1	2·8	2·5	2·2	1·9	1·5
8	8·3	7·9	7·6	7·3	6·9	6·6	6·3	6·0	5·6	5·3	5·0	4·6	4·3	4·0	3·6	3·3	3·0	2·6	2·3	2·0	1·7
9	8·8	8·5	8·1	7·8	7·4	7·1	6·7	6·3	6·0	5·6	5·3	4·9	4·6	4·2	3·9	3·5	3·2	2·8	2·5	2·1	1·8
10	9·4	9·0	8·6	8·3	7·9	7·5	7·1	6·8	6·4	6·0	5·6	5·3	4·9	4·5	4·1	3·8	3·4	3·0	2·6	2·3	1·9
11	10·0	9·6	9·2	8·8	8·4	8·0	7·6	7·2	6·8	6·4	6·0	5·6	5·2	4·8	4·4	4·0	3·6	3·2	2·8	2·4	2·0
12	10·7	10·2	9·8	9·4	9·0	8·5	8·1	7·7	7·2	6·8	6·4	6·0	5·5	5·1	4·7	4·3	3·8	3·4	3·0	2·6	2·1
13	11·3	10·9	10·4	10·0	9·5	9·1	8·6	8·2	7·7	7·3	6·8	6·4	5·9	5·4	5·0	4·5	4·1	3·6	3·2	2·7	2·3
14	12·1	11·6	11·1	10·6	10·1	9·7	9·2	8·7	8·2	7·7	7·2	6·8	6·3	5·8	5·3	4·8	4·3	3·9	3·4	2·9	2·4
15	12·8	12·3	11·8	11·3	10·8	10·3	9·7	9·2	8·7	8·2	7·7	7·2	6·7	6·2	5·6	5·1	4·6	4·1	3·6	3·1	2·6
16	13·6	13·1	12·5	12·0	11·4	10·9	10·4	9·8	9·3	8·7	8·2	7·6	7·1	6·5	6·0	5·4	4·9	4·4	3·8	3·3	2·7
17	14·5	13·9	13·3	12·7	12·2	11·6	11·0	10·4	9·8	9·3	8·7	8·1	7·5	6·9	6·4	5·8	5·2	4·6	4·1	3·5	2·9
18	15·4	14·7	14·1	13·5	12·9	12·3	11·7	11·1	10·4	9·8	9·2	8·6	8·0	7·4	6·8	6·1	5·5	4·9	4·3	3·7	3·1
19	16·3	15·6	15·0	14·3	13·7	13·0	12·4	11·7	11·1	10·4	9·8	9·1	8·5	7·8	7·2	6·5	5·9	5·2	4·6	3·9	3·3
20	17·3	16·6	15·9	15·2	14·5	13·8	13·1	12·4	11·8	11·1	10·4	9·7	9·0	8·3	7·6	6·9	6·2	5·5	4·8	4·1	3·5
21	18·3	17·6	16·9	16·1	15·4	14·7	13·9	13·2	12·5	11·7	11·0	10·3	9·5	8·8	8·1	7·3	6·6	5·9	5·1	4·4	3·7
22	19·4	18·6	17·9	17·1	16·3	15·5	14·8	14·0	13·2	12·4	11·6	10·9	10·1	9·3	8·5	7·8	7·0	6·2	5·4	4·7	3·9
23	20·6	19·7	18·9	18·1	17·3	16·5	15·6	14·8	14·0	13·2	12·3	11·5	10·7	9·9	9·0	8·2	7·4	6·6	5·8	4·9	4·1
24	21·8	20·9	20·0	19·2	18·3	17·4	16·5	15·7	14·8	13·9	13·1	12·2	11·3	10·4	9·6	8·7	7·8	7·0	6·1	5·2	4·4
25	23·0	22·1	21·2	20·3	19·3	18·4	17·5	16·6	15·7	14·7	13·8	12·9	12·0	11·1	10·1	9·2	8·3	7·4	6·4	5·5	4·6
26	24·4	23·4	22·4	21·4	20·5	19·5	18·5	17·5	16·6	15·6	14·6	13·6	12·7	11·7	10·7	9·7	8·8	7·8	6·8	5·8	4·9
27	25·8	24·7	23·7	22·7	21·6	20·6	19·6	18·5	17·5	16·5	15·5	14·4	13·4	12·4	11·3	10·3	9·3	8·2	7·2	6·2	5·2
28	27·2	26·1	25·0	23·9	22·9	21·8	20·7	19·6	18·5	17·4	16·3	15·2	14·2	13·1	12·0	10·9	9·8	8·7	7·6	6·5	5·4
29	28·7	27·6	26·4	25·3	24·1	23·0	21·8	20·7	19·5	18·4	17·2	16·1	14·9	13·8	12·6	11·5	10·3	9·2	8·0	6·9	5·7
30	30·3	29·1	27·9	26·7	25·5	24·3	23·1	21·8	20·6	19·4	18·2	17·0	15·8	14·6	13·4	12·1	10·9	9·7	8·5	7·3	6·1
31	32·0	30·7	29·5	28·2	26·9	25·6	24·3	23·1	21·8	20·5	19·2	17·9	16·7	15·4	14·1	12·8	11·5	10·2	9·0	7·7	6·4
32	33·8	32·4	31·1	29·7	28·4	27·0	25·7	24·3	23·0	21·6	20·3	18·9	17·6	16·2	14·9	13·5	12·2	10·8	9·5	8·1	6·8
33	35·6	34·2	32·8	31·4	29·9	28·5	27·1	25·7	24·2	22·8	21·4	20·0	18·5	17·1	15·7	14·3	12·8	11·4	10·0	8·6	7·1
34	37·6	36·0	34·5	33·0	31·5	30·0	28·5	27·0	25·5	24·0	22·5	21·0	19·5	18·0	16·5	15·0	13·5	12·0	10·5	9·0	7·5
35	39·6	38·0	36·4	34·8	33·2	31·7	30·1	28·5	26·9	25·3	23·7	22·2	20·6	19·0	17·4	15·8	14·2	12·7	11·1	9·5	7·9
36	41·7	40·0	38·3	36·7	35·0	33·3	31·7	30·0	28·3	26·7	25·0	23·3	21·7	20·0	18·3	16·7	15·0	13·3	11·7	10·0	8·3
37	43·9	42·1	40·4	38·6	36·9	35·1	33·3	31·6	29·8	28·1	26·3	24·6	22·8	21·1	19·3	17·5	15·8	14·0	12·3	10·5	8·8
38	46·2	44·3	42·5	40·6	38·8	36·9	35·1	33·2	31·4	29·5	27·7	25·9	24·0	22·2	20·3	18·5	16·6	14·8	12·9	11·1	9·2

Appendix: Electronic Circuits and Sources of Supply of Components

I. Electronic Components

A. Resistors

Figure 58 shows the symbols used in circuit diagrams to represent resistors and potentiometers. Small fixed resistors have their values indicated by a series of coloured bands, usually known as the "resistor colour code" (Table 11). With only six exceptions in Chapter 2, the fixed resistors used in circuits in this book are 0.5 W, ± 5% tolerance carbon film resistors. The exceptions are the 6R8 resistor in the circuit of the silicon solar cell solarimeter, which should be of the wirewound type, with a power rating of 2.5 W (or greater than 2.5 W), and the five resistors marked with an asterisk on the millivoltmeter circuit, which should be ± 2% metal oxide.

Potentiometers and larger resistors usually have their values indicated numerically, but it is now almost universal practice to avoid the use of decimal points (because of the danger of the points becoming obliterated) and to use the multiplying factor in their place. Thus, 1 000 000 ohm (1 megohm) would be shown as 1 M, and 1 500 000 ohm (1.5 megohm) resistor would be 1M5. Similarly, a 2700 ohm resistor (2.7 kilohms) would be 2k7 and a 6.8 ohm resistor would be 6R8. This convention has been followed in the circuit diagrams in this book.

A potentiometer is a resistor fitted with a moving contact so that a proportion of the voltage developed across the resistor can be selected. A variable resistor (or rheostat) is a potentiometer with only the moving contact and one of the fixed connections used. If the arrow used in the circuit symbol is sharp, then a "front panel" control is indicated, and a blunt arrow indicates a "preset", or "screwdriver-adjustment", control. Pre-set potentiometers are available with wirewound, carbon or "cermet" (metallised ceramic) resistive

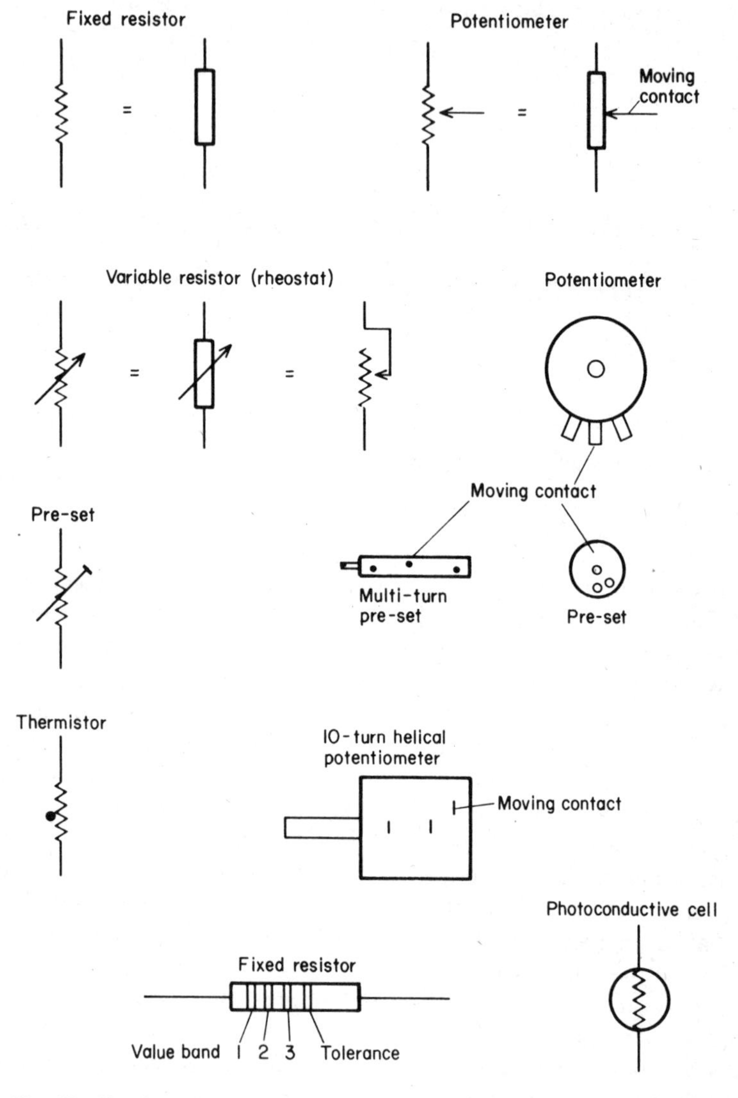

Fig. 58. Circuit symbols for resistors and the pin connections of potentiometers.

Table 11

Resistor colour code. Small fixed resistors have their resistances specified by a system of coloured bands to an accuracy of two significant figures. The first two bands specify a two-digit number, which is multiplied by a factor given by the third band. The fourth band indicates the tolerance.

Colour of band	Interpretation for band: 1	2	3	4
Silver	–	–	× 0.01	± 10%
Gold	–	–	× 0.1	± 5%
Black	0	0	× 1	–
Brown	1	1	× 10	–
Red	2	2	× 100	± 2%
Orange	3	3	× 1000	–
Yellow	4	4	× 10 000	–
Green	5	5	× 100 000	–
Blue	6	6	× 1 000 000	–
Violet	7	7	–	–
Grey	8	8	–	–
White	9	9	–	–

Preferred values
Resistors are normally made with values in a series in which each value is 20% greater than the preceding one, e.g. 10, 12, 15, 18, 22, 27, 33, 39, 47, 56, 68, 82, 100.

elements. Single-turn cermet pre-set potentiometers are suitable for use in any of the circuits in this book, except where multi-turn pre-set potentiometers are specified in the text. Multi-turn cermet pre-sets have a linear cermet track but use a screw thread to control the moving contact. The 10-turn helical potentiometers specified as front panel controls in some circuits are a larger and more elaborate device with a helical resistance track. This arrangement gives the potentiometer a high degree of stability and accuracy. The pin connections of the common types of potentiometer are shown in Fig. 58.

All types of resistor, including thermistors and photoconductive cells, can be connected either way round (polarity is not important). The same is true of potentiometers, provided the moving contact is correctly identified, but if the control is found to work "backwards", then the leads to the two fixed contacts should be reversed.

B. Capacitors

There are two types of capacitor used in the circuits in this book. Those in

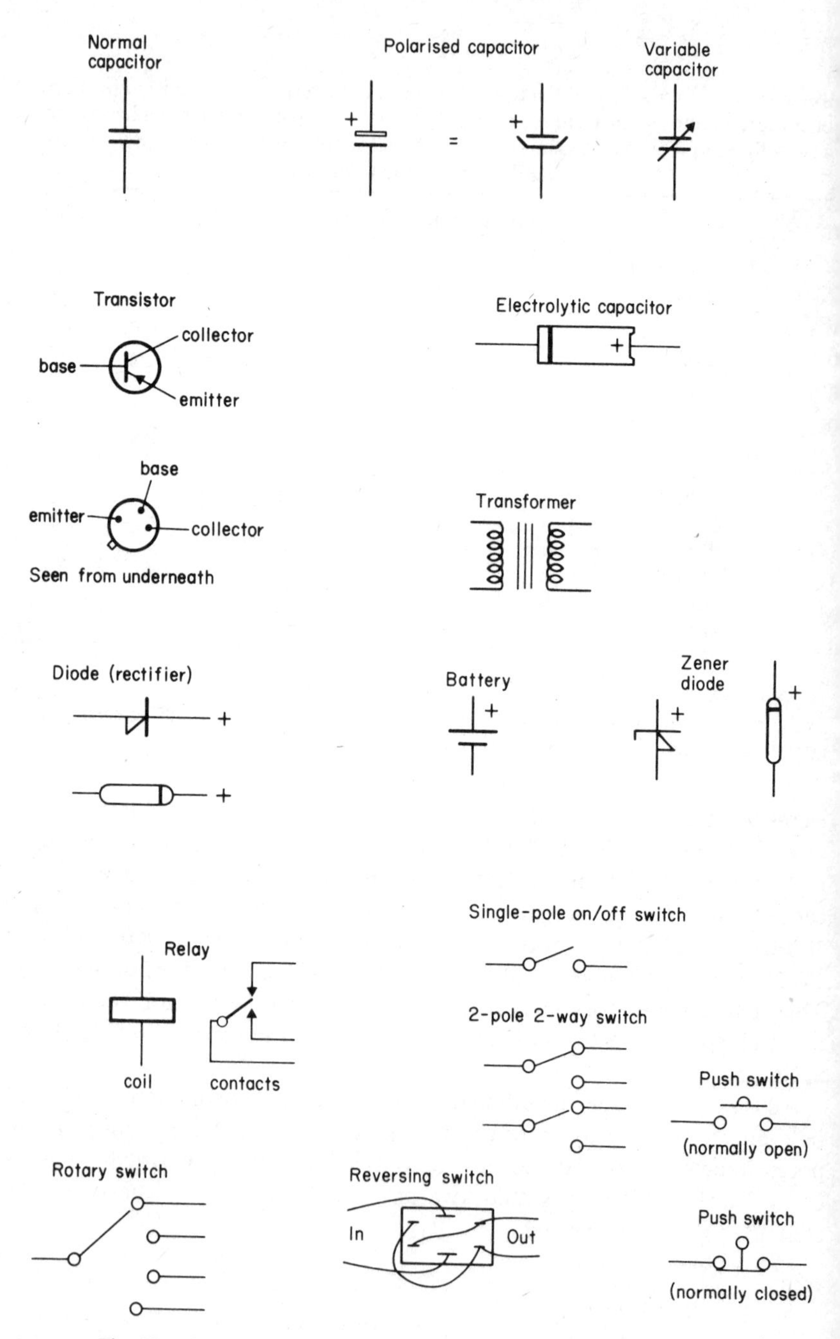

Fig. 59. Circuit symbols and connections of other discrete components.

which the two sides are shown as black bars (normal capacitor in Fig. 59) which can be connected either way round, are polyester or polycarbonate types. These have a voltage rating of between 63 and 400 volts, and since the highest voltage used in the circuits in this book is 27 V, any of them will be satisfactory. The capacitors shown with one of the bars of the symbol black and the other in outline are polarised capacitors (Fig. 59) and the polarity must be observed. All the polarised capacitors in this book are electrolytic capacitors, with a voltage rating of 16 V or above, with the exception of the 220 μF capacitor in the circuit of the cup anemometer indicator in Chapter 4, which should have a rating of at least 40 V. The positive terminal of electrolytic capacitors is usually indicated with a plus sign, and the negative with a black band. In cases of doubt, the lead connected to the aluminium can is negative.

The values of capacitances in all the circuits are given in microfarads.

C. *Diodes and Transistors*

The circuit symbols representing diodes and transistors, and how these relate to the physical device are shown in Fig. 59. The only type of transistor used in this book is the 2N3053, but any medium-power "NPN" transistor would work equally well (NPN indicates the polarity of the transistor, and the opposite type is PNP, which would not be suitable). Ordinary diodes are specified as 1N914, but any small general-purpose diode, such as 1N4001 s eries, 0A202 or 1N4148 are suitable. Zener diodes, which are voltage stabilising devices, have their voltages specified in a similar way to the values of small fixed resistors, either using the resistor colour code to specify the rating in millivolts (e.g. yellow, purple and red, which gives 4700 would indicate 4.7 V), or numerically, using V in place of the decimal point (e.g. 6V8 indicates a 6.8 V device). Any 0.5 or 1 W zener diode, such as the BZY88 series, can be used in the circuits in this book. The polarity of all diodes must be observed, and it should be noted that unlike electrolytic capacitors, the black band on diodes indicates the positive lead.

D. *Integrated-circuit Operational Amplifier*

Figure 60 shows the circuit symbol, connections and typical application of an integrated-circuit operational amplifier. The integrated circuits used in this book are the 741 (R.S. Components Ltd, 305-311) general-purpose operational amplifier or a high-impedance version with field-effect transistor (FET) input (R.S. Components Ltd, 307-058), which has identical pin connections. The 741 is equivalent to the SN72741P and the FET version is equivalent to the μAF355TC.

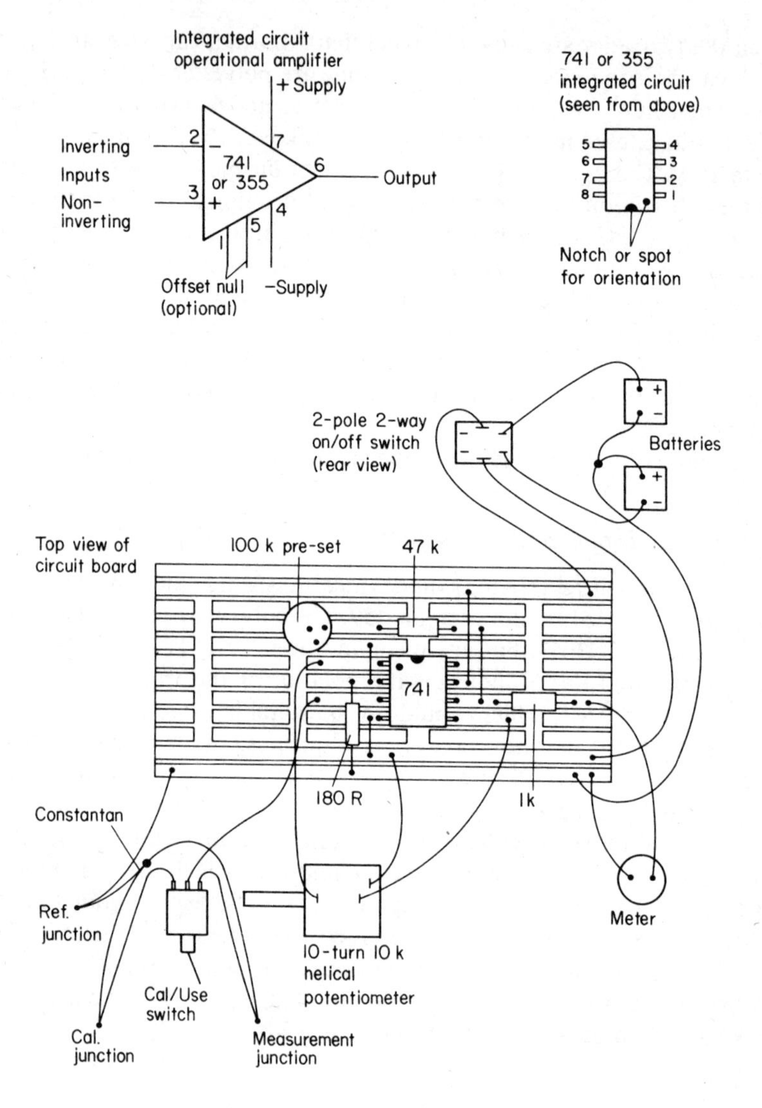

Fig. 60. Integrated-circuit operational amplifier symbol, pin connections and application.

The example of the use of the operational amplifier in Fig. 60 is taken from the circuit of the thermocouple indicator in Chapter 1. The integrated-circuit is mounted on a piece of stripboard (R.S. Components Ltd, 433-949) which consists of strips of copper, broken at intervals to allow the connection of integrated circuits, on a substrate of glass-fibre/epoxy resin. Some continuous copper strips are also provided to act as power supply rails, and the whole board is perforated with 1 mm holes on a 1 mm × 1 mm matrix. The copper strips run on the lower side of the board, and the upper side is usually printed to show the distribution of the strips. The components are mounted only on the upper side, and their leads are soldered to the copper strips on the lower side. For clarity, Fig. 60 shows holes only where they are used. The 741 operational amplifier may be soldered directly into the board, but sockets are available, known as 8-way DIL sockets (R.S. Components Ltd, 401-683), which can be soldered to the board, and into which the integrated circuit can be inserted.

E. Other Components and Materials

The transformer used in the oscillator circuit for the bridge and indicator in Chapter 3 is a small device intended for driving the output stage in a transistor radio (R.S. Components Ltd, 217-624). Other small audio transformers which have a tapped winding may be suitable, but it could be necessary to change the value of the 0.22 μF capacitor connected across the transformer and possibly the values of the two 4k7 resistors that connect the transformer to the bridge.

The silicon solar cell used in the solarimeter circuit in Chapter 2 (R.S. Components Ltd, 307-137) measures about 29 × 23 mm and produces 80 mA at an input of 1 kW/m^2. Any solar cell of about the same dimensions should be satisfactory.

The cadmium sulphide photoconductive cell used in Chapter 2 (ORP12, R.S. Components Ltd, 305-620), which is sometimes described as a light-dependent resistor, is a general-purpose device which has been used in applications ranging from flame-failure indication in boilers to the brilliance control of television receivers. It has a typical cell resistance of 2k4 at 50 lux and 130R at 1000 lux, and any cadmium sulphide cell of about the same size (14 mm diameter) should work equally well.

The thermistors used in the designs in this book all have a resistance of about 5000Ω at 20°C. In the prototypes a miniature type was used (R.S. Components Ltd, 151-142) but any type or size of thermistor with a similar resistance can be used.

The silicon diode used as a temperature sensor in Chapter 1 is the 1N914 computer diode, but if a smaller diode is available, it can be used with advantage.

The voltage regulators used in a number of circuits can be any small regulator with a current rating of 100 mA or above (R.S. Components Ltd, 306-190 for 5 volts and 306-207 for 12 volts). The precision voltage reference (ZN423) used in the silicon diode temperature meters in Chapter 1 (R.S.

Components Ltd, 283-233) is actually an integrated circuit, although it is used in the same way as a zener diode. These devices are often used as a reference in digital voltmeters, and may be available as a spare part for such instruments.

The digital panel meter used in two designs in Chapter 1, but which could be used for almost any design in the book with a little modification, has a range of 199.9 mV and a resolution of 0.1 mV. R.S. Components Ltd have a version with liquid crystal display (258-827) or with light-emitting diodes (258-811). The latter is easier to see in the dark, but uses very much more power. Many such meters are now available, and it may also be convenient to use a portable digital multi-meter, which can be switched to read voltage, resistance or current.

The analogue meters (100 μA and 1 mA) used in the prototypes had internal resistances of 1300 Ω (R.S. Components Ltd, 259-606) and 75 Ω (259-612) respectively. In many of the circuits the actual value of the internal resistance is not important, but the silicon solar cell solarimeter will need some adjustment of resistance values if a meter with a different resistance is used. The same may also be true of the microbalance in Chapter 5, and the CdS light meter.

The reed switch used in the cup anemometer (R.S. Components Ltd, 338-147) has a glass portion 20 mm long and 3.2 mm in diameter. Similar reed switches are available encapsulated in a plastic block for use in burglar alarm systems.

The electromechanical counter used in the cup anemometer (R.S. Components Ltd, 259-892) is a six-digit reset counter rated at 24 V, and any similar device should work, with some small adjustments to the pulse length being required for slower counters. An alternative approach for those who do not wish to build circuits, is a new device known as an electronic totaliser (R.S. Components Ltd, 258-798). Although rather an expensive component, it can be connected directly to the reed switch, with few external components. Data is supplied with the device.

The aerosol cans of refrigerant used for dew-point determination (Chapter 3) are intended by the makers for tracing cracks in printed-circuit boards (R.S. Components Ltd, 554-765). The same refrigerant is available for use in blowing dust from optical surfaces in the photographic trade, but in this case it is necessary to invert the can to get the liquid refrigerant.

It would not be helpful to show the pin connections of relays, because of the large variation, but it is common practice to print a diagram of the connections on the case of the device. Rotary switches also show great variation in design, and the only advice which can be offered here is that an ohm-meter, or a battery and bulb connected in series, can be used to investigate how the switch works.

II. Tools

Many electronic components are small, and it is best to use the smallest cutters and pliers which can be obtained. A soldering iron with a replaceable bit is useful, particularly if thermocouples are to be made. The thermopile solarimeter in Chapter 2 makes considerable demands on both soldering iron and operator, and a very small bit is a great asset in this type of work.

Some of the designs in this book require very small holes to be drilled. While these holes can be made with a hand drill, a miniature electric drill is much easier to control. R.S. Components Ltd sell a very useful 12 V drill (543-945), which can be powered from a battery, and which can be used with drills up to 1.6 mm diameter (543-951 for a set). A set of accessories including a wire brush, a polisher, a slitting saw and a pair of reamers is also available for this drill (543-697). A similar drill and accessories are available from Radio Shack in the United States.

There is a great deal to be said in favour of plastic instrument cases, particularly for field equipment. Those without access to a good workshop will also find that they are easy to work with simple hand tools. The prototypes of the instruments described in this book were all made in moulded polystyrene cases (R.S. Components Ltd, 509-282 for the solar cell solarimeters, 509-591 for the cup anemometer, and the bridge/indicator in Chapter 3, and 509-579 for the rest).

III. Sources of Supply of Components and Materials

Most of the components used in the designs in this book were supplied by R.S. Components Ltd, who are wholesale suppliers to industry, the radio retail trade and scientific establishments. They have the following offices in the United Kingdom:

Head Office: R S Components Ltd
P O Box 427
13-17 Epworth Street
London EC2P 2HA
Telephone 01-253 3040 (day and night)
01-250 3131 (day)
Telex: 262341

Export enquiries Telephone 01-253 1222

Midlands Office:	R S Components Ltd P O Box 253 Saltley Trading Estate Birmingham B8 1BQ Telephone 021-328 0233 (day and night) Telex: 337294
North West Office	R S Components Ltd P O Box 12 Kennedy Way, Green Lane Trading Estate Stockport Cheshire SK4 2JT Telephone 061-477 8400 (day and night) Telex: 666610

The Tandy Corporation, which sells in the United States and Canada under the name of Radio Shack, claim to have over 7000 branches around the world, and can also supply most of the electronic components needed to construct the instruments described in this book.
They have the following main offices:

USA:	Radio Shack Fort Worth Texas 76102
Canada	Radio Shack Barrie Ontario L4M 4W5
Australia	Tandy Corporation 280-316 Victoria Road Rydalmere New South Wales 2116
Belgium:	Tandy Corporation Parc Industriel de Nannine 5140 Nannine
United Kingdom:	Tandy Corporation Branch UK Bilston Road Wednesbury West Midlands WS10 7JN Telephone 021-556 6101

Thermocouple materials, such as copper and constantan cables and connectors can be obtained from:

T C Ltd
21 Windsor Road
Uxbridge UB8 1AD
Telephone 0895 36822/35971

References

Cloudsley-Thompson, J. L. (1967). "Microecology." Edward Arnold, London.

Corbet, S. A., Willmer, P. G., Beament, J. W. L., Unwin, D. M. and Prŷs-Jones, O. E. (1979a). Post-secretory determinants of sugar concentration in nectar. *Plant, Cell and Environment* **2**, 293 – 308.

Corbet, S. A., Unwin, D. M. and Prŷs-Jones, O. E. (1979b). Humidity, nectar and insect visits to flowers, with special reference to *Crataegus, Tilia* and *Echium*. *Ecological Entomology* **4**, 9 – 22.

Easty, A. C. and Young, S. (1976). A small scale dewpoint humidity measurer. *Journal of Physics E* **9**, 106 – 110.

Geiger, R. (1965). "The Climate near the Ground." Harvard University Press, Cambridge, Mass. and London.

Lee, R. (1969). Chemical temperature integration. *Journal of Applied Meteorology* **8**, 432-430.

McPherson, H. G. (1969). Photocell-filter combinations for measuring photo-synthetically active radiation. *Agricultural Meteorology,* **6**, 347 – 356.

Meteorological Office (1964a). "Hygrometric Tables." Part II. Her Majesty's Stationery Office, London.

Meteorological Office (1964b). "Hygrometric Tables." Part III. Her Majesty's Stationery Office, London.

Miall, R. C. (1978). The flicker fusion frequency of six laboratory insects, and the response of the compound eye to mains fluorescent "ripple". *Physiological Entomology* **3**, 99 – 106.

Monteith, J. L. (1972). "Survey of Instruments for Micrometeorology."
Blackwell Scientific Publications, Oxford and Edinburgh.

Monteith, J. L. (1973). "Principles of Environmental Physics." Edward Arnold, London.

Rosenberg, N. J. (1974). "Microclimate: The Biological Environment." Wiley, New York and London.

Schwerdtfeger, P. (1976). "Physical Principles of Micro-meteorological Measurements." Elsevier, Amsterdam, Oxford and New York.

Solomon, M. E. (1951). Control of humidity with potassium hydroxide, sulphuric acid and other solutions. *Bulletin of Entomological Research* **42**, 543 – 554.

Stoutjesdijk, P. (1974). The open shade, an interesting microclimate. *Acta Botanica Neerlandica* **23**, 125 – 130.
Szeicz, G. (1966). Field measurements in the 0.4 – 0.7 micron range. In "Light as an Ecological Factor" (R. Bainbridge, G.C. Evans and O. Rackham, eds), 41 – 52. Blackwell Scientific Publications, Oxford.
Unwin, D. M. (1978). Simple techniques for microclimate measurement. *Journal of Biological Education* **12**, 179 – 189.
Unwin, D.M. (1979). Details of the miniature thermocouple psychrometer. In Corbet *et al*, 1979b, p 22.
Van Praagh, J. P. (1975). Light-ripple and visual acuity in a climate room for honey-bees (*Apis mellifera* L.). *Netherlands Journal of Zoology* **25**, 506 – 515.
Weast, R. C. (1979). "Handbook of Chemistry and Physics." 49th Edn. Chemical Rubber Company, Cleveland, Ohio.
Williams, G. D. V. and Brochu, J. (1969). "Vapour pressure deficit vs. relative humidity for expressing atmospheric moisture content." *Naturaliste Canadien* **96**, 621 – 636.
Winston, P. W. and Bates, H. B. (1960). Saturated solutions for the control of humidity in biological research. *Ecology* **41**, 232 – 237.

Subject index